Eva Hackmann

Geodesic equations in General Relativity

Eva Hackmann

Geodesic equations in General Relativity

Motion in black hole space-times with and without cosmological constant

Südwestdeutscher Verlag für Hochschulschriften

Impressum/Imprint (nur für Deutschland/ only for Germany)
Bibliografische Information der Deutschen Nationalbibliothek: Die Deutsche Nationalbibliothek verzeichnet diese Publikation in der Deutschen Nationalbibliografie; detaillierte bibliografische Daten sind im Internet über http://dnb.d-nb.de abrufbar.

Alle in diesem Buch genannten Marken und Produktnamen unterliegen warenzeichen-, marken- oder patentrechtlichem Schutz bzw. sind Warenzeichen oder eingetragene Warenzeichen der jeweiligen Inhaber. Die Wiedergabe von Marken, Produktnamen, Gebrauchsnamen, Handelsnamen, Warenbezeichnungen u.s.w. in diesem Werk berechtigt auch ohne besondere Kennzeichnung nicht zu der Annahme, dass solche Namen im Sinne der Warenzeichen- und Markenschutzgesetzgebung als frei zu betrachten wären und daher von jedermann benutzt werden dürften.

Verlag: Südwestdeutscher Verlag für Hochschulschriften Aktiengesellschaft & Co. KG
Dudweiler Landstr. 99, 66123 Saarbrücken, Deutschland
Telefon +49 681 37 20 271-1, Telefax +49 681 37 20 271-0
Email: info@svh-verlag.de
Zugl.: Bremen, Universität, Diss., 2010

Herstellung in Deutschland:
Schaltungsdienst Lange o.H.G., Berlin
Books on Demand GmbH, Norderstedt
Reha GmbH, Saarbrücken
Amazon Distribution GmbH, Leipzig
ISBN: 978-3-8381-1799-7

Imprint (only for USA, GB)
Bibliographic information published by the Deutsche Nationalbibliothek: The Deutsche Nationalbibliothek lists this publication in the Deutsche Nationalbibliografie; detailed bibliographic data are available in the Internet at http://dnb.d-nb.de.

Any brand names and product names mentioned in this book are subject to trademark, brand or patent protection and are trademarks or registered trademarks of their respective holders. The use of brand names, product names, common names, trade names, product descriptions etc. even without a particular marking in this works is in no way to be construed to mean that such names may be regarded as unrestricted in respect of trademark and brand protection legislation and could thus be used by anyone.

Publisher: Südwestdeutscher Verlag für Hochschulschriften Aktiengesellschaft & Co. KG
Dudweiler Landstr. 99, 66123 Saarbrücken, Germany
Phone +49 681 37 20 271-1, Fax +49 681 37 20 271-0
Email: info@svh-verlag.de

Printed in the U.S.A.
Printed in the U.K. by (see last page)
ISBN: 978-3-8381-1799-7

Copyright © 2010 by the author and Südwestdeutscher Verlag für Hochschulschriften Aktiengesellschaft & Co. KG and licensors
All rights reserved. Saarbrücken 2010

Acknowledgment

First of all I would like to thank my supervisor Prof. Dr. Claus Lämmerzahl for providing me with this interesting topic, his good advice, and for always having time to help despite his considerable workload. Without our numerous interesting, detailed, and helpful discussions and his physical as well as formal explanatory information this book would not exist. Also, I am indebted to Prof. Dr. Peter H. Richter for his interest in my research, his comprehensive introduction to hyperelliptic functions, and for directing my attention to some most useful papers.

A thanks goes also to Dennis Lorek, Dr. Holger Dullin, Prof. Dr. Wolfgang Fischer, Prof. Dr. Hansjörg Dittus, Prof. Dr. Jutta Kunz, and in particular to Dr. Valeria Kagramanova for fruitful discussions and the good collaboration.

Likewise I want to express my gratitude to my office colleagues Andreas Resch for answering my computer questions and Dr. Thorben Könemann for the hardware support.

Further I am much obliged to Prof. Dr. Hans Rath for the opportunity to work at the Center of Applied Space Technology and Microgravity and to my colleagues of the fundamental physics group for their support and the good time.

The financial support of the German Research Foundation DFG is gratefully acknowledged.

Many thanks are also given to the Südwestdeutscher Verlag für Hochschulschriften for publishing this book and for the good cooperation.

Finally, I am deeply grateful to my family for their constant support over all these years, to all my friends for their help, and to Daniel for everything.

Contents

1 **Introduction**   1

2 **Mathematical preliminaries**   6

   2.1 Weierstrass functions 7

       2.1.1 The Weierstrass $\wp$ function 8

       2.1.2 The Weierstrass ζ and σ functions 10

   2.2 Differentials on Riemann surfaces 12

   2.3 Theta functions 15

   2.4 Solution methods 17

3 **Geodesics in spherically symmetric space-times**   21

   3.1 General types of geodesics 23

   3.2 Schwarzschild and Reissner-Nordström space-times 26

       3.2.1 Geodesics in Schwarzschild space-times 26

       3.2.2 Geodesics in Reissner-Nordström space-times 32

   3.3 Schwarzschild- and Reissner-Nordström-
(anti-)de Sitter space-times 38

       3.3.1 Geodesics in Schwarzschild-(anti-)de Sitter space-times 39

Contents

    3.3.2   Geodesics in Reissner-Nordström-(anti-)de Sitter space-times 56

3.4 Higher-dimensional space-times 63

4 Geodesics in axially symmetric space-times   67

4.1 General types of orbits 69

4.2 Kerr space-time 73

    4.2.1   Types of orbits 74

    4.2.2   Analytical solutions of the geodesic equations 85

4.3 Kerr-de Sitter space-time 90

    4.3.1   Types of orbits 92

    4.3.2   Analytical solution of geodesic equations 105

    4.3.3   Analytic expressions for observables 114

4.4 Plebański-Demiański space-times 116

5 Summary and Outlook   120

5.1 Summary 120

5.2 Outlook 122

A Elliptic integrals of third kind   125

A.1 General solution procedure 126

A.2 Post-Schwarzschild periastron advance 129

B Calculational details   134

B.1 Typical integral expressions 134

B.2 Calculation of periods 136

B.3 Calculation of redundant parameter 137

CHAPTER 1

Introduction

The motions of stars and planets were observed by mankind for thousands of years, sparked their interest, and excited their inquisitiveness. Although early highly developed cultures already had a certain knowledge about the motion of stars, the modern scientific approach is mainly based on the works of Johannes Kepler (1571-1630), Isaac Newton (1643-1727), and Albert Einstein (1879-1955).

In 1609, Kepler postulated in his work *Astronomica Nova* the existence of a force radiated by the Sun, which decreases with distance and which causes the planets to move faster if closer to the Sun. Based on this assumption and the analysis of the orbital data of Mars, he found that the planets move on elliptical rather than circular orbits and developed the first and second law of planetary motion later named after him. Kepler published his third law in 1618, which formulated the connection between the length of the semi-major axis of a planet and its orbital period. That enabled him to correctly compute the orbital velocity of a planet. With these three laws, Kepler became one of the founders of modern astronomy.

However, it was not before 1687, when Newton's *Philosophiae Naturalis Principia Mathematica* was published, that the force acting over a distance postulated by Kepler was interpreted as *Gravitation* (the word was introduced by Newton) caused by the mass of the planet. Newton unified the laws of Kepler as effects of the gravitational force, which he found to be proportional to the inverse square of the distance to the central object as formulated in his law of gravitation. His work also distinguishes no longer between the laws of motion on the earth and in space, which was a basic scientific insight.

Using Newton's gravitational law the motion of planets could not only be predicted, but fundamentally explained. This included not only the gravitational force of the Sun, but also the much

1. Introduction

weaker influence of the planets on each other. These disturbances of the perfect elliptic planetary orbits were most evident in the motion of the periapsis of the innermost planet Mercury, an effect known as perihelion advance. In 1859, the mathematician Le Verrier was able to compute the perihelion advance of Mercury using Newton's law with a result of approximately 530 arcseconds per century, which was considerably lower than the observed value. This was the first hint that Newton's theory of gravitation was not able to fully explain the motion of planets although at that time some unknown solar system objects rather than an incomplete theory were suspected.

Only with the introduction of General Relativity by Albert Einstein this discrepancy could be consistently explained. Einstein's theory defined gravitation no longer as a force acting over a distance but as a geometrical curvature of space and time, which is described by Einstein's field equations and which causes the motion of particles and light in space-time. In 1915, the gap between the observed and Le Verrier's predicted value of the perihelion shift of Mercury could be explained as relativistic effect [1] using slightly wrong field equations with an approximated solution for a static and isotropic space-time describing the gravitational field of the Sun. A year later, Einstein found the correct field equations and Karl Schwarzschild was able to formulate an exact static and isotropic solution of these field equations [2], which was named after him. With these achievements, the unexplained additional perihelion advance of Mercury could finally be confirmed to be a relativistic effect. Together with the observation of the deflection of light by Dyson, Eddington, and Davidson in 1919 [3] this resulted in the final breakthrough of General Relativity.

For a thorough understanding of the physical properties of solutions of Einstein's field equations it is essential to study the orbits of test particles and light rays in these space-times. On the one hand, this is important from an observational point of view, since only matter and light are observed and, thus, can give insight into the physics of a given gravitational field. On the other hand, this study is also important from a fundamental point of view, since the motion of matter and light can be used to classify a given space-time, to decode its structure and to highlight its characteristics. Indeed, it can be shown that the space-time geometry can be constructed from the concepts of light propagation and freely falling test particles [4, 5]. After the discovery of the static and spherically symmetric vacuum solution of Einstein's field equations by Schwarzschild the motion of particles and light described by the geodesic equation were studied in this space-time. In 1931 Hagihara presented the analytical solution of the geodesic equation in Schwarzschild space-time using the theory of elliptic functions [6]. He also extensively discussed the complete set of timelike as well as null geodesics. The same methods can be used to solve the geodesic equation in Reissner-Nordström space-time [7, 8], an exact solution of the gravitational field equations derived in 1918 including an electrical charge of the gravitating object.

However, already in 1918 Lense and Thirring found that the rotation of gravitating objects in General Relativity has an additional effect on the geodesics not known in Newton's theory [9], which has to be taken into account for the many rotating astronomical objects. The exact solution of a stationary and axially symmetric space-time describing such a rotation was presented by Roy Patrick Kerr in 1963 [10]. Around 1970, the structure of this more complicated space-time was investigated in many aspects by Carter [11, 12], who was also able to demonstrate the separability of the Hamilton-Jacobi equations, which led to a fourth constant of motion, the Carter constant, and ensured the integrability of the geodesic equation. In the sixties and seventies a number of publications dealt with the motion of particles and light in Kerr space-time, for example de Felice [13] who extensively treated equatorial orbits. Their results have been collected, reviewed, and extended by Chandrasekhar in 1983 [14]. A complete elaboration of all types of geodesics in Kerr space-time including their analytical expressions has only be carried through recently by Slezáková [15].

In 1917, soon after the formulation of General Relativity, Einstein considered the consequences of his theory not only for objects in the solar system, but for the universe as a whole. He found that his field equations do not allow the universe to be static but that it has to expand or collapse [16], which contradicted widely spread and, in particular, his own beliefs. As an ad-hoc hypothese, he introduced a small cosmological constant in the field equations, which should enable a static universe without changing anything on solar system scales. However, in 1927 Lemaître inferred from the known distances to some galaxies, published by Hubble shortly before, and the redshift of their spectral lines that the universe is expanding, and gave a value of the Hubble parameter near the one found by Hubble himself in 1929. Therefore, the cosmological constant as introduced by Einstein lost its significance but was nevertheless discussed in some models of the universe.

Even though the idea of a non-zero cosmological constant was commonly rejected after 1929, it attracted again wide interest in 1998 with the observation of distant, high redshifted supernovae. The data indicates that the universe is not only expanding but does so at an accelerated rate [17, 18]. To explain the expansion behavior of the universe and related observations, like the fluctuations in the cosmic microwave background or structure formation, dark energy was introduced [19, 20], which can be modeled (among other possibilities as quintessence [21, 22]) as a very small positive cosmological constant representing a constant energy density homogeneously distributed in space-time. As a consequence, it is necessary in principle to describe all observations related to gravity within a framework including the cosmological constant. However, it has been shown within an approximation scheme based on the frame given by the Schwarzschild-de Sitter space-time that the cosmological constant plays no role in all solar system observations or in strong field effects [23, 24]. Also within the rotating version of this solution, the Kerr-de Sitter solution, no observable effects arise [25].

Nevertheless, there have been some discussions on whether the Pioneer anomaly, the unexplained acceleration of the Pioneer 10 and 11 spacecraft toward the inner solar system of $a_{\text{Pioneer}} = (8.47 \pm 1.33) \times 10^{-10}$ m/s^2 [26] which is of the order of cH where H is the Hubble constant, may be related to the cosmological expansion and, thus, to the cosmological constant. The same order of acceleration is present also in the galactic rotation curves which astonishingly can be successfully modeled using a modified Newtonian dynamics involving an acceleration parameter a_MOND which again is of the order of 10^{-10} m/s^2. This mysterious coincidence of characteristic accelerations appearing at different scales and the fact that all these phenomena appear in a weak gravity or weak acceleration regime raise the question whether the linear approximation schemes used in [23, 24, 25] really hold. Therefore, it is desireable to obtain analytical solutions of the equations of motion for a definite answer of these questions.

There have been also some discussions if the cosmological constant has a measureable effect on the physics of binary systems, which play an important role in testing General Relativity. Although such an effect would be very small, it could influence the creation of gravitational waves [27, 28]. In particular, the observation of gravitational waves originating from extreme mass ratio inspirals (EMRIs) is a main goal of the Laser Interferometer Space Antenna (LISA). The calculation of such gravitational waves benefits from analytical solutions of geodesic equations not only by improved accuracy, which is, in principle, arbitrary high, but also by the prospect of developing fast semi-analytically computing methods [29]. Also, analytical solutions offer a systematic approach to determine the last stable spherical and circular orbits, which are starting points for inspirals and used for the calculation of gravitational wave templates.

Analytical solutions are especially useful for the analysis of the properties of a space-time not only from an academic point of view. In fact, they offer a frame for tests of the accuracy and reliability of numerical integrations due to their, in principle, unlimited accuracy. In addition, they can be used to systematically calculate all observables in the given space-time with the very high accuracy needed for the understanding of some observations.

In this book, the analytical solutions of the geodesic equations in Schwarzschild, Reissner-Nordström, and Kerr space-time will be generalized to the case of a nonvanishing cosmological constant. The geodesics in these space-times with and without cosmological constant will be classified and compared. Furthermore, the analytic expressions for observables will be given in each space-time.

The book starts with a chapter about the mathematical tools which will be needed to derive the analytic expressions for the mentioned space-times. The theory of elliptic functions used by Hagihara [6] for the analytical solution of the geodesic equation in Schwarzschild space-time will be presented as far as necessary. Afterwards, the theory of hyperelliptic functions, which contains the elliptic functions as a special case and which will be used in the following to analytically solve the geodesic equations in space-times with nonvanishing cosmological constant, will be outlined. This includes the concept of Riemann surfaces, the definition of theta functions, and the solution of Jacobi's inversion problem.

Subsequently, geodesics in spherically symmetric and static space-times, namely the Schwarzschild and Reissner-Nordström space-times as well as their generalizations with a nonvanishing cosmological constant will be discussed in the third chapter. The analytical solutions of the geodesic equations in Schwarzschild and Reissner-Nordström space-time will be rederived and the set of all geodesics is classified according to the energy and angular momentum of the test particle or light ray. However, the focus of this chapter lies on geodesics in the corresponding space-times with nonvanishing cosmological constant Λ. The analytical solution of the geodesic equations will be elaborated and the resulting geodesics will be compared to the case of a vanishing cosmological constant. Analytic expressions for the perihelion shift and its series expansion with respect to Λ will be given. The analytical solution of the geodesic equation in Schwarzschild-de Sitter space-time is also applied to the question whether the cosmological constant might be the origin of the anomalous acceleration of the Pioneer spacecraft. It is also noted that the methods presented in this chapter can be applied to higher-dimensional spherically symmetric space-times.

In chapter 4 the methods used in the foregoing chapters are generalized and applied to the case of axially symmetric space-times, namely to Kerr and Kerr-de Sitter space-time. In both space-times, the equations of motion are derived depending on the proper time and decoupled following an idea of Mino [30]. Then possible types of test particle orbits are classified and discussed focussing on the influence of the cosmological constant. For a nonvanishing cosmological constant we derive the analytic expressions for some observables of particle and light trajectories. For bound orbits, the periastron advance and the Lense-Thirring effect are given in terms of the fundamental orbital frequencies. Finally, the solution method demonstrated for Kerr-de Sitter space-time is shown to be applicable to the general class of Plebański-Demiański space-times without acceleration, which contains all space-times with separable Hamilton-Jacobi equation.

CHAPTER 2

Mathematical preliminaries

In this chapter we will review and explain the mathematical tools needed for an analytical solution of the geodesic equations in the spherically and axially symmetric black hole space-times considered in this book. Although the geodesic equation is in general a set of partial differential equations, the symmetries in the space-times considered here allow in most cases a reduction to ordinary differential equations of the form

$$\left(y^i \frac{dy}{dx}\right)^2 = P(y), \tag{2.0.1}$$

where P is a polynomial of degree 6 or lower and $i = 0$ or $i = 1$. For the case that P is of degree 1 or 2 the solution of such an equation can be given in terms of elementary functions [31] whereas for higher degrees in general elliptic or hyperelliptic functions are needed. A differential equation (2.0.1) is said to be elliptic or of elliptic type if P is of degree 3 or 4 with only simple zeros and hyperelliptic or of hyperelliptic type if P is of degree 5 or higher with only simple zeros. The theory of these functions was developed already in the 19th century mainly by Jacobi [32], Abel [33], Riemann [34, 35], and Weierstrass [36]. An extended review of their achievements was given in a seminal book by Baker [37] exposing the whole theory in a compact way. In 1931 Hagihara used elliptic functions to analytically solve the geodesic equation in Schwarzschild space-time, but afterwards the interest faded out and although the theory of elliptic functions is still a standard part of function theory, the knowledge on hyperelliptic functions is a very special subject. Only in the past years it attracted again attention in the theory of solitons, see [38, 39] and references within. Recently, the theory of elliptic and hyperelliptic functions was again used to analytically solve geodesic equations [40, 41, 42].

In the following we will present the elliptic and hyperelliptic functions used in the other parts of this book to solve geodesic equations. In particular, we will deduce all mathematical tools needed to analytically solve the geodesic equations resulting in differential equations of type (2.0.1) where P is of degree 5 or 6, namely the geodesic equations in Schwarzschild-de Sitter, Kerr-de Sitter, general Plebański-Demiański without acceleration, and higher-dimensional static and spherically symmetric space-times.

First, the Weierstrass functions are reviewed. All differential equations of elliptic type in this book are solved in terms of these functions. The analogous expressions in terms of Jacobi elliptic functions are completely omitted here but can at least partly be found in [15, 42]. After introducing the notion of Riemann surfaces, theta functions are reviewed which form the basis of the Kleinian sigma functions used to solve all geodesic equations of hyperelliptic type. In this context also the solution of Jacobi's inversion problem, which is considered in the last section, is of importance.

2.1 Weierstrass functions

The general term *Weierstrass functions* is used to denote a special elliptic function constructed in the most simple way possible together with two associated non-elliptic functions. Thus, before defining any of these the concept of elliptic functions themselves has to be explained. Shortly said, they are the analogue of the trigonometric functions with two instead of one period.

Definition 2.1. *1. A function $f : \mathbb{C} \to \mathbb{C}$ is meromorphic in $D \subset \mathbb{C}$ if f is holomorphic in $D \backslash S_f$, where $S_f \subset D$ is a discrete set, and has no essential singularities in S_f.*

2. Let f be meromorphic in $\mathbb{C}$ and c a constant. If $f(z+c) = f(z)$ for all $z \in \mathbb{C}$ then c is a period of f and f is called periodic.

3. A meromorphic function f is elliptic (or doubly periodic) if its set of all periods is given by $\{2m\omega + 2n\omega' \mid m, n \in \mathbb{Z}\}$ with the periods 2ω, $2\omega'$, where $\frac{\omega}{\omega'} \notin \mathbb{R}$.

The System of all elliptic functions with periods $2\omega, 2\omega'$ is denoted by $K(2\omega, 2\omega')$. The constants ω and ω' are also called half periods.

In the following chapters we will need some basic facts about elliptic functions. The proofs are omitted here as they can be found in many books on the subject, for example [43, 44, 45].

Theorem 2.2. *Let f be an elliptic function. Then:*

1. If f has no poles then f is a constant.

2. Mathematical preliminaries

2. *f possesses a finite number of poles in each period parallelogram $\{z_0 + 2t_1\omega + 2t_2\omega' \mid 0 \leq t_1, t_2 < 1\}$, where $z_0 \in \mathbb{C}$ and ω, ω' are the half periods.*

3. *For a non-constant f the sum of all residuals in the period parallelogram is zero.*

The period parallelogram $\{2t_1\omega + 2t_2\omega' \mid 0 \leq t_1, t_2 < 1\}$ is called fundamental. An example of such a fundamental period parallelogram can be found in Fig. 3.14. The set of all elliptic functions with half periods (ω, ω') is a field with respect to the addition and multiplication of functions, which also contains all derivatives of its elements. In particular, if f is an elliptic function with half periods (ω, ω') then $R(f)$, where R is a rational function, is also an elliptic function with the same half periods (ω, ω').

2.1.1 The Weierstrass $\wp$ function

From Thm. 2.2 it follows that the most simple non-constant elliptic function has two simple poles with residuals 1 and -1 or a double pole with residual 0. The Weierstrass elliptic $\wp$ function introduced in [36] realizes the latter possibility.

Definition 2.3. *The Weierstrass $\wp$-function is defined by*

$$\wp(z) := \frac{1}{z^2} + \sum_{n,m \in \mathbb{Z}, (n,m) \neq 0} \left(\frac{1}{(z - z_{nm})^2} - \frac{1}{z_{nm}^2} \right), \qquad (2.1.1)$$

where $z_{nm} = 2n\omega + 2m\omega'$ with the two half periods ω and ω'.

The Weierstrass elliptic $\wp$ function has some useful properties which are also frequently needed for solving geodesic equations.

1. Counted with multiplicity, $\wp$ takes every value in the period parallelogram exactly two times.

2. It is $\wp(u) = \wp(v)$ for $u, v \in \mathbb{C}$ if and only if $u - v$ or $u + v$ is a period of $\wp$.

3. Thus, $\wp(v)$ has multiplicity 2 iff $2v$ is a period of $\wp$. Thus, in the fundamental period parallelogram the points of multiplicity 2 are $v = 0$, $v = \omega$, $v = \omega'$, and $v = \omega + \omega'$.

4. The zeros of $\wp'$ are given by $e_3 := \wp(\omega)$, $e_2 := \wp(\omega + \omega')$, and $e_1 := \wp(\omega')$.

An important feature of $\wp$ is that it solves an easy differential equation.

2.1. Weierstrass functions

Theorem 2.4. *The Weierstrass elliptic $\wp$ function solves the differential equation*

$$(\wp'(u))^2 = 4(\wp(u))^3 - g_2\wp(u) - g_3, \tag{2.1.2}$$

where the invariants g_2, g_3 are given by

$$g_2 = 60 \sum_{m,n\in\mathbb{Z},(m,n)\neq 0} \frac{1}{z_{mn}^4}, \quad g_3 = 140 \sum_{m,n\in\mathbb{Z},(m,n)\neq 0} \frac{1}{z_{mn}^6}.$$

Sometimes the invariants g_2, g_3 are explicitly mentioned in the argument of the $\wp$ function, i.e. $\wp(z) \equiv \wp(z; g_2, g_3)$, if it is not clear to which periods or invariants the $\wp$ function corresponds. Note that by the theorems of Vieta the invariants g_2, g_3 are connected to the zeros e_1, e_2, e_3 of $\wp'$ (which are the zeros of the polynomial $4y^3 - g_2y - g_3$) by

$$\begin{aligned} e_1 + e_2 + e_3 &= 0 \\ -4(e_1e_2 + e_2e_3 + e_3e_1) &= g_2 \\ 4e_1e_2e_3 &= g_3 \end{aligned} \tag{2.1.3}$$

The connection between the zeros of $\wp'$ and the periods $2\omega, 2\omega'$ of $\wp$ (and $\wp'$) is even more pronounced. The periods $2\omega, 2\omega'$ can be calculated by

$$\omega = \int_{e_3}^{\infty} \frac{dz}{\sqrt{4z^3 - g_2z - g_3}}, \quad \omega' = \int_{-\infty}^{e_1} \frac{dz}{\sqrt{4z^3 - g_2z - g_3}}. \tag{2.1.4}$$

In this book, the Weierstrass $\wp$ function is most often used to solve the initial value problem

$$\left(\frac{dy}{dx}\right)^2 = 4y^3 - g_2y - g_3, \quad y(x_0) = y_0 \tag{2.1.5}$$

in the following way: On the one hand the differential equation (2.1.2) implies that

$$x = \int_0^x dz = \int_{\wp(0)}^{\wp(x)} \frac{d\wp(z)}{\wp'(z)} = \int_{\infty}^{\wp(x)} \frac{dy}{\sqrt{4y^3 - g_2y - g_3}}, \tag{2.1.6}$$

where the sign of the square root has to be chosen to be identical with the sign of $\wp'$. On the other hand a separation of variables in (2.1.5) yields

$$\begin{aligned} x - x_0 &= \int_{y_0}^{y} \frac{dz}{\sqrt{4z^3 - g_2z - g_3}} \\ \Leftrightarrow x - x_{in} &= \int_{\infty}^{y} \frac{dz}{\sqrt{4z^3 - g_2z - g_3}}, \end{aligned} \tag{2.1.7}$$

where $x_{in} = x_0 + \int_{y_0}^{\infty} \frac{dz}{\sqrt{4z^3 - g_2z - g_3}}$ depends only on initial values. Therefore, (2.1.5) can be solved by

$$y = \wp(x - x_{in}). \tag{2.1.8}$$

9

2. Mathematical preliminaries

Even a more general problem

$$\left(\frac{dz}{dx}\right)^2 = P(z) = \sum_{i=0}^{4} a_i z^i, \quad z(x_0) = z_0, \tag{2.1.9}$$

where P is a polynomial of degree 3 or 4 with only simple zeros, can be solved in this way by applying up to two substitutions. If P is of degree 4 the substitution $z = \xi^{-1} + z_P$, where z_P is a zero of P, transforms the problem to

$$\left(\frac{d\xi}{dx}\right)^2 = P_3(\xi) = \sum_{i=0}^{3} b_i \xi^i, \quad \xi(x_0) = \xi_0 \tag{2.1.10}$$

with a polynomial P_3 of degree 3. Subsequently, or if P is a polynomial of degree 3 in the first place, a substitution $\xi = \frac{1}{b_3}\left(4y - \frac{b_2}{3}\right)$ (or $z = \frac{1}{a_3}\left(4y - \frac{a_2}{3}\right)$, respectively) casts the problem in the form (2.1.5) with

$$\begin{aligned} g_2 &= \frac{1}{16}\left(\frac{4}{3}b_2^2 - 4b_1 b_3\right), \\ g_3 &= \frac{1}{16}\left(\frac{1}{3}b_1 b_2 b_3 - \frac{2}{27}b_2^3 - b_0 b_3^2\right), \end{aligned} \tag{2.1.11}$$

(or with b_i replaced by a_i, respectively), which can then be solved by (2.1.8). Differential equations of the type (2.1.9) are called elliptic differential equations of first kind.

2.1.2 The Weierstrass ζ and σ functions

The geodesic equation considered in this book are sometimes even more general than the problems (2.0.1) or (2.1.5). For these kinds of equations the Weierstrass ζ and σ functions are helpful. The ζ function is defined by

$$\frac{d}{dz}\zeta(z) = -\wp(z), \quad \lim_{z \to 0}\left(\zeta(z) - \frac{1}{z}\right) = 0$$

which implies through

$$\zeta(z) - \frac{1}{z} = -\int_0^z \wp(z') - \frac{1}{z'^2}\,dz' \tag{2.1.12}$$

(the integration path runs along an arbitrary continuous and piecewise differentiable curve from 0 to z in the fundamental period parallelogram) the equation

$$\zeta(z) = \frac{1}{z} + \sum_{m,n \in \mathbb{Z}, (m,n) \neq 0}\left(\frac{1}{z - z_{mn}} + \frac{1}{z_{mn}} + \frac{z}{z_{mn}^2}\right). \tag{2.1.13}$$

From this expression it is already obvious that ζ is a meromorphic function with simple poles z_{nm}. In addition, ζ is an odd function, i.e. $\zeta(-z) = -\zeta(z)$. The condition $\frac{d}{dz}\zeta(z) = -\wp(z)$ gives

$$-\frac{d}{du}\left(\zeta(u + 2\omega) - \zeta(u)\right) = \wp(u + 2\omega) - \wp(u) = 0.$$

2.1. Weierstrass functions

Therefore, $\zeta(u+2\omega) - \zeta(u) =: 2\eta$ is constant. Analogously a constant $2\eta' = \zeta(u+2\omega') - \zeta(u)$ can be defined. These constant are also known as the periods of second kind. There is a simple relation between the half periods ω, ω' and the periods of second kind, the Legendre relation

$$\frac{1}{2}\pi i = \omega'\eta - \omega\eta'. \qquad (2.1.14)$$

Finally, the Weierstrass σ function is defined by

$$\frac{d}{dz}\log\sigma(z) = \frac{\sigma'(z)}{\sigma(z)} = \zeta(z), \quad \lim_{z\to 0}\frac{\sigma(z)}{z} = 1 \qquad (2.1.15)$$

and, thus,

$$\sigma(z) = z\exp\left(\int_0^z \zeta(z') - \frac{1}{z'}dz'\right) = z\prod_{m,n\in\mathbb{Z},(m,n)\neq 0}\left(1 - \frac{z}{z_{mn}}\right)\exp\left(\frac{z}{z_{mn}} + \frac{z^2}{2z_{mn}^2}\right).$$

From this equation it is obvious that σ is an odd function. Note that we will sometimes denote the Weierstrass sigma function with $\sigma^{(W)}$ to avoid confusions with the Kleinian sigma function, which will be introduced in Sec. 2.4. Let us analyse now how σ changes if a period $2\omega, 2\omega'$ of $\wp$ is added to its argument. We integrate the expression

$$\frac{\sigma'(z+2\omega)}{\sigma(z+2\omega)} - \frac{\sigma'(z)}{\sigma(z)} = \zeta(z+2\omega) - \zeta(z) = 2\eta$$

and obtain

$$\log\frac{\sigma(z+2\omega)}{\sigma(z)} = 2\eta z + c \quad \text{or} \quad \sigma(z+2\omega) = \sigma(z)\exp(2\eta z + c).$$

The constant of integration c can be determined by setting $z = -\omega$ which gives

$$\sigma(\omega) = -\sigma(\omega)e^{-2\eta\omega+c} \Rightarrow c = \pi i + 2\eta\omega + 2\pi i k, \ k\in\mathbb{Z}.$$

It follows that the σ function fulfills the relation

$$\sigma(z+2\omega) = -\sigma(z)e^{2\eta(z+\omega)}. \qquad (2.1.16)$$

In this book, the Weierstrass ζ and σ functions are used to solve elliptic integrals of the type

$$\int_{y_1}^{y_2}\frac{f(y)\,dy}{\sqrt{4y^3 - g_2 y - g_3}}, \qquad (2.1.17)$$

where f is a rational function. With a substitution $y = \wp(v)$ this integral can be transformed to

$$\int_{v_1}^{v_2}\frac{f(\wp(v))\,\wp'(v)dv}{\sqrt{4\wp(v)^3 - g_2\wp(v) - g_3}} = \int_{v_1}^{v_2} f(\wp(v))dv, \qquad (2.1.18)$$

where the sign of the square root has to be chosen according to the sign of $\wp'$. The function $F(v) := f(\wp(v))$ is elliptic with the half periods ω, ω' of $\wp$ and can be expressed in terms of the Weierstrass ζ function and its derivatives as the following theorem, which can be found, e.g., in [43], shows.

2. Mathematical preliminaries

Theorem 2.5. *Let F be an arbitrary elliptic function with poles p_i of multiplicity m_i in the fundamental period parallelogram, i.e. the meromorphic part of F near p_i is given by*

$$\sum_{n=1}^{m_i}(-1)^{n-1}\frac{(n-1)!A_i^{n-1}}{(z-p_i)^n} \qquad (2.1.19)$$

for some constants A_i^{n-1}. Then F can be written as

$$F(z) = c + \sum_{p_i}\left(\sum_{n=0}^{m_i-1} A_i^n \frac{d^n}{dz^n}\zeta(z-p_i)\right) \qquad (2.1.20)$$

$$= c + \sum_{p_i}\left(A_i^0\zeta(z-p_i) - A_i^1\wp(z-p_i) - \sum_{j=1}^{m_i-2} A_i^{j+1}\wp^{(j)}(z-p_i)\right),$$

where the first sum runs over all poles p_i and c is a constant.

With this theorem, the elliptic integral (2.1.17) can be integrated to

$$\int_{y_1}^{y_2}\frac{f(y)dy}{\sqrt{4y^3 - g_2 y - g_3}} = c(v_2 - v_1) + \sum_{p_i}\left[A_i^0(\log\sigma(v_2-p_i) - \log\sigma(v_1-p_i))\right.$$

$$\left. + A_i^1(\zeta(v_2-p_i) - \zeta(v_1-p_i)) - \sum_{j=0}^{m_i-3} A_i^{j+2}(\wp^{(j)}(v_2-p_i) - \wp^{(j)}(v_1-p_i))\right], \qquad (2.1.21)$$

where p_i are the poles of multiplicity m_i of $F(v) := f(\wp(v))$ and $\wp(v_i) = y_i$.

The definitions and basic properties presented in this section are already sufficient to solve all geodesic equations of elliptic type in this book. In the remaining sections of this chapter the mathematical foundations for solving geodesic equations of hyperelliptic type are described.

2.2 Differentials on Riemann surfaces

Let X be the compact Riemann surface of the algebraic function $x \mapsto \sqrt{P_n(x)}$ for a polynomial P_n of degree n. It can be represented as the algebraic curve

$$X := \{z = (x,y) \in \mathbb{C}^2 \,|\, y^2 = P_n(x)\} \qquad (2.2.1)$$

[46] or as the analytic continuation of $\sqrt{P_n}$. The last one can be realized as a two-sheeted covering of the Riemann sphere which can be constructed in the following way: let e_i, $i = 1, \ldots, n$, be the zeros of P_n and, in the case that n is odd, $e_{n+1} = \infty$. These are the so-called branch points. Now take two copies of the Riemann sphere, one for each of the two possible values of $\sqrt{P_n}$, and cut them between every two branch points (e_i, e_{i+1}) in such a way that the cuts do not touch each other. These are the

2.2. Differentials on Riemann surfaces

so-called branch cuts, see Fig. 2.1. Of course, the two copies have to be identified at the branch points where the two values of $\sqrt{P_n}$ are identical. They are then glued together along the branch cuts in such a way that $\sqrt{P_n}$ together with all its analytic continuations is uniquely defined on the whole surface. On this surface $x \mapsto \sqrt{P_n(x)}$ is now a single-valued function. This construction can be visualized as a "pretzel", see Fig. 2.1. For a strict mathematical description of the construction of a compact Riemann surface, see [47], for example.

Compact Riemann surfaces are characterized by their genus g. This can be defined as the dimension of the space of holomorphic differentials on the Riemann surface or, topologically seen, as the number of 'holes' of the Riemann surface. If P_n has only simple zeros the genus of the Riemann surface X of $\sqrt{P_n}$ is equal to $g = \left[\frac{n-1}{2}\right]$, where $[x]$ denotes the largest integer less than or equal to x [48]. Thus, if $n = 5$ or $n = 6$ as it is the case for the geodesic equations of hyperelliptic type considered in this book the genus of the Riemann surface is $g = 2$. Every compact Riemann surface of genus g can be equipped with a homology basis $\{a_i, b_i \,|\, i = 1, \ldots, g\} \in H_1(X, \mathbb{Z})$ of closed paths as shown in Fig. 2.1.

In order to construct periodic functions we first have to define a canonical basis of the space of holomorphic differentials $\{dz_i \,|\, i = 1, \ldots, g\}$ and of associated meromorphic differentials $\{dr_i \,|\, i = 1, \ldots, g\}$ on the Riemann surface by

$$dz_i := \frac{x^{i-1} dx}{\sqrt{P_n(x)}}, \tag{2.2.2}$$

$$dr_i := \sum_{k=i}^{2g+1-i} (k+1-i) b_{k+1+i} \frac{x^k dx}{4\sqrt{P_n(x)}}, \tag{2.2.3}$$

with b_j being the coefficients of the polynomial $P_n(x) = \sum_{j=1}^{n} b_j x^j$ [48]. We also introduce the period matrices $(2\omega, 2\omega')$ and $(2\eta, 2\eta')$ related to the homology basis

$$\begin{aligned} 2\omega_{ij} &:= \oint_{a_j} dz_i, & 2\omega'_{ij} &:= \oint_{b_j} dz_i, \\ 2\eta_{ij} &:= -\oint_{a_j} dr_i, & 2\eta'_{ij} &:= -\oint_{b_j} dr_i. \end{aligned} \tag{2.2.4}$$

The differentials in (2.2.2) and (2.2.3) have been chosen such that the components of the related period matrices fulfill the Legendre relation

$$\begin{pmatrix} \omega & \omega' \\ \eta & \eta' \end{pmatrix} \begin{pmatrix} 0 & -\mathbb{1}_g \\ \mathbb{1}_g & 0 \end{pmatrix} \begin{pmatrix} \omega & \omega' \\ \eta & \eta' \end{pmatrix}^t = -\frac{1}{2}\pi i \begin{pmatrix} 0 & -\mathbb{1}_g \\ \mathbb{1}_g & 0 \end{pmatrix}, \tag{2.2.5}$$

where $\mathbb{1}_g$ is the $g \times g$ unit matrix, [48]. Note that Eq. (2.1.14) for the periods of elliptic functions is the special case of $g = 1$ in (2.2.5).

2. Mathematical preliminaries

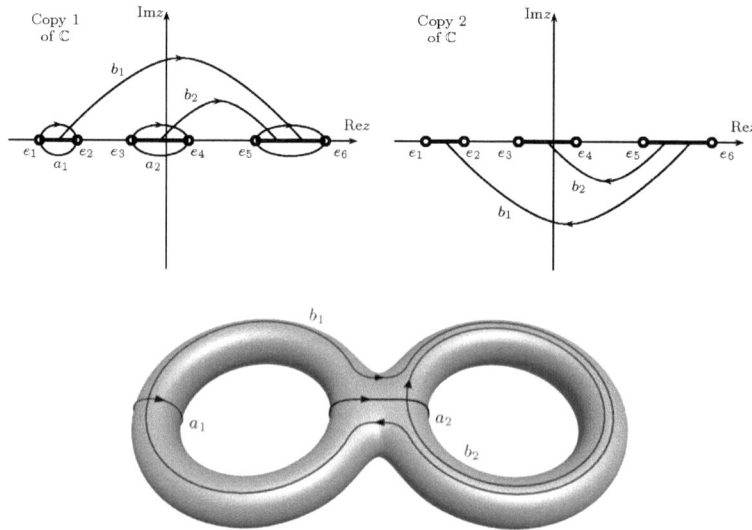

Figure 2.1: Riemann surface of genus $g = 2$ with real branch points $e_1, \ldots, e_6$. Upper figure: Two copies of the complex plane with closed paths giving a homology basis $\{a_i, b_i \,|\, i = 1, \ldots, g\}$. The branch cuts (thick solid lines) are chosen from e_{2i-1} to e_{2i}, $i = 1, \ldots, g+1$. Lower figure: The "pretzel" with the topologically equivalent homology basis.

Also, we introduce the normalized holomorphic differentials

$$d v := (2\omega)^{-1} dz, \qquad dz = (dz_1, dz_2, \ldots, dz_g)^t. \qquad (2.2.6)$$

The period matrix related to these differentials is given by $(\mathbb{1}_g, \tau)$, where τ is defined by

$$\tau := \omega^{-1} \omega'. \qquad (2.2.7)$$

It can be shown [49] that this normalized matrix τ always is a Riemann matrix, that is, τ is symmetric and its imaginary part is positive definite.

In this book, the holomorphic differentials introduced above are most often used as a convenient way to formulate geodesic equations of the type

$$\left(y^i \frac{dy}{dx} \right)^2 = P_n(y), \quad y(x_0) = y_0, \qquad (2.2.8)$$

where $i = 0$ or $i = 1$ and $n = 5$. Indeed, an integration of (2.2.8) yields

$$x - x_0 = \int_{y_0}^{y} \frac{y^i dy}{\sqrt{P_n(y)}} = \int_{y_0}^{y} dz_i \qquad (2.2.9)$$

with the definition (2.2.2) of dz_i.

However, not all geodesic equations of hyperelliptic type can be formulated like (2.2.8). For the solution of such equations, in addition the canonical differential of third kind is needed, which has two simple poles poles x_1, x_2 with vanishing total residual. The most simple construction of a differential of third kind is

$$dP(x_1, x_2) = \left(\frac{y + y_1}{x - x_1} - \frac{y + y_2}{x - x_2} \right) \frac{dx}{2\sqrt{P_n(x)}}, \qquad (2.2.10)$$

which has simple poles in the points (x_1, y_1) and (x_2, y_2) of the Riemann surface represented by the algebraic curve $\{(x, y) \in \mathbb{C}^2 \,|\, y^2 = P_n(x)\}$ with residuals $+1$ and -1, respectively (cf. [48, 50]).

The definition (2.2.10) can now be used to reformulate hyperelliptic integrals of the form

$$\int_{y_1}^{y_2} \frac{dy}{(y - t)\sqrt{P_n(y)}}. \qquad (2.2.11)$$

From Eq. (2.2.10) it can be inferred that

$$dP(t^+, t^-) = \left(\frac{y + \sqrt{P_n(t)}}{x - t} - \frac{y - \sqrt{P_n(t)}}{x - t} \right) \frac{dx}{2\sqrt{P_n(x)}} = \sqrt{P_n(t)} \frac{dx}{(x - t)\sqrt{P_n(x)}}, \qquad (2.2.12)$$

where $t^+ := (t, +\sqrt{P_n(t)})$ and $t^- := (t, -\sqrt{P_n(t)})$ and, thus,

$$\int_{y_1}^{y_2} \frac{dy}{(y - t)\sqrt{P_n(y)}} = \frac{1}{\sqrt{P_n(t)}} \int_{y_1}^{y_2} dP(t^+, t^-). \qquad (2.2.13)$$

2.3 Theta functions

A compact Riemann surface of genus g has $2g$ independent closed paths, given by the homology basis $\{a_i, b_i \,|\, i = 1, \ldots, g\}$ and each corresponding to a period of the functions related to these surfaces. In order to construct $2g$-periodic functions, we need the theta function $\vartheta : \mathbb{C}^g \to \mathbb{C}$,

$$\vartheta(z; \tau) := \sum_{m \in \mathbb{Z}^g} e^{i\pi m^t (\tau m + 2z)}. \qquad (2.3.1)$$

The series on the right-hand side converges absolutely and uniformly on compact sets in $\mathbb{C}^g$ and, thus, defines a holomorphic function in $\mathbb{C}^g$. This is obvious from the estimate $\text{Re}(m^t(i\tau)m^t) \leq -cm^t m$ for a constant $c > 0$, what follows from the fact that $\text{Re}(i\tau)$ is negative definite. The theta function

2. Mathematical preliminaries

is periodic with respect to the columns of $\mathbb{1}_g$ and quasiperiodic with respect to the columns of τ, i.e., for any $n \in \mathbb{Z}^g$ the relations

$$\vartheta(z + \mathbb{1}_g n; \tau) = \vartheta(z; \tau), \tag{2.3.2}$$

$$\vartheta(z + \tau n; \tau) = e^{-i\pi n^t(\tau n + 2z)} \vartheta(z; \tau) \tag{2.3.3}$$

hold. In the following chapters, the theta functions with characteristics $g, h \in \frac{1}{2}\mathbb{Z}^g$ will also be needed[1], which are defined by

$$\vartheta[g, h](z; \tau) := \sum_{m \in \mathbb{Z}^g} e^{i\pi(m+g)^t(\tau(m+g)+2z+2h)}$$

$$= e^{i\pi g^t(\tau g + 2z + 2h)} \vartheta(z + \tau g + h; \tau). \tag{2.3.4}$$

Later it will be important that for every g, h the set $\Theta_{\tau g + h} := \{z \in \mathbb{C}^g \mid \vartheta[g, h](z; \tau) = 0\}$, called a *theta divisor*, is a $(g-1)$-dimensional subset of the Jacobian $\operatorname{Jac}(X)$ of the Riemann surface X with Riemann matrix τ, see [49].

Some functions closely related to the theta function will be needed in Sec. 2.4 to formulate the solution of Jacobi's inversion problem. First, consider the Riemann theta function

$$\vartheta_e(x; \tau) := \vartheta\left(\int_{x_0}^{x} dv - e; \tau\right), \tag{2.3.5}$$

with some arbitrary but fixed $e \in \mathbb{C}^g$. The Riemann vanishing theorem, see e.g. [49], states that the Riemann theta function is either identically to zero or has exactly g zeros $x_1, \ldots, x_g$ for which

$$\sum_{i=1}^{g} \int_{x_0}^{x_i} dv = e + K_{x_0} \tag{2.3.6}$$

holds (modulo periods). Here $K_{x_0} \in \mathbb{C}^g$ is the vector of Riemann constants with respect to the base point x_0 given by

$$K_{x_0, j} = \frac{1 + \tau_{jj}}{2} - \sum_{l \neq j} \oint_{a_l} \left(\int_{x_0}^{x} dv_j\right) dv_l(x) \tag{2.3.7}$$

(where τ_{jj} is the jth diagonal element). If the base point x_0 is equal to ∞, this vector can be determined by

$$K_\infty = \sum_{i=1}^{g} \int_{\infty}^{e_{2i}} dv, \tag{2.3.8}$$

where e_{2i} is the starting point of one of the branch cuts not containing ∞ for each i, see [48]. Hence, K_∞ can be expressed as a linear combination of half periods in this case. For problems of hyperelliptic

---
[1]The symbol $\frac{1}{2}\mathbb{Z}^g$ denotes the set of all g-dimensional vectors with integer or half integer entries $\ldots, -\frac{3}{2}, -1, -\frac{1}{2}, 0, \frac{1}{2}, 1, \frac{3}{2}, \ldots$

nature it is usually assumed that the Riemann theta function ϑ_e does not vanish identically. However, here we are interested in the opposite case: in Sec. 2.4 Jacobi's inversion problem will be restricted to the set of zeros of $\vartheta(\,\cdot\, + K_{x_0}; \tau)$, i.e. to the theta divisor $\Theta_{K_{x_0}}$. This application of the theta divisor was first used in [51] for the case of a double pendulum.

2.4 Solution methods

Let us consider now the Abel map

$$\mathcal{A}_{x_0} : X \to \mathrm{Jac}(X), \quad x \mapsto \int_{x_0}^{x} dz \qquad (2.4.1)$$

from the Riemann surface X to the Jacobian $\mathrm{Jac}(X) = \mathbb{C}^g/\Gamma$ of X, where $\Gamma = \{2\omega v + 2\omega' v' \mid v, v' \in \mathbb{Z}^g\}$ is the lattice of periods of the differential $dz = (dz_1, \ldots, dz_g)^t$. The image $\mathcal{A}_{x_0}(X)$ of X by this continuous function is of complex dimension one and, thus, an inverse map $\mathcal{A}_{x_0}^{-1}$ is not defined for all points of $\mathrm{Jac}(X)$. However, the g-dimensional Abel map

$$A_{x_0} : S^g X \to \mathrm{Jac}(X), \quad (x_1, \ldots, x_g)^t \mapsto \sum_{i=1}^{g} \int_{x_0}^{x_i} dz \qquad (2.4.2)$$

from the gth symmetric product $S^g X$ of X (the set of unordered 'vectors' $(x_1, \ldots, x_g)^t$ where $x_i \in X$) to the Jacobian is one-to-one almost everywhere. Jacobi's inversion problem is now to determine $x \in S^g(X)$ for given $y \in \mathrm{Jac}(X)$ from the equation

$$y = A_{x_0}(x). \qquad (2.4.3)$$

In the case of $g = 2$, which appears in all geodesic equations of hyperelliptic type in this book, this reads using the definition (2.2.2) of $dz = (dz_1, dz_2)$

$$\begin{aligned} y_1 &= \int_{x_0}^{x_1} \frac{dx}{\sqrt{P_n(x)}} + \int_{x_0}^{x_2} \frac{dx}{\sqrt{P_n(x)}}, \\ y_2 &= \int_{x_0}^{x_1} \frac{x\,dx}{\sqrt{P_n(x)}} + \int_{x_0}^{x_2} \frac{x\,dx}{\sqrt{P_n(x)}}, \end{aligned} \qquad (2.4.4)$$

where $n = 5$ or $n = 6$ is the degree of the polynomial P_n. We will see later, that the geodesic equations of hyperelliptic type in this book can be solved as a limiting case of Jacobi's inversion problem.

The solution of (2.4.3) can be formulated in terms of the derivatives of the Kleinian sigma function $\sigma : \mathbb{C}^g \to \mathbb{C}$ defined by

$$\sigma(z) = C e^{-\frac{1}{2} z^t \eta \omega^{-1} z} \vartheta[g, h]\left((2\omega)^{-1} z; \tau\right), \qquad (2.4.5)$$

2. Mathematical preliminaries

where the constant C can be given explicitly, see [48], but does not matter here. To avoid confusions with the Weierstrass sigma function introduced in Sec. 2.1, we will denote the Kleinian sigma function by $\sigma^{(K)}$ and the Weierstrass sigma function by $\sigma^{(W)}$ if necessary. Furthermore, we will write $\sigma(z; \tau)$, $\sigma(z; g, h)$, $\sigma(z; x)$, where $x = \tau g + h$, or some combination of this if the definition of σ is not clear. Note that z is a zero of the Kleinian sigma function if and only if $(2\omega)^{-1}z$ is a zero of the theta function $\theta[g, h]$. The second logarithmic derivatives of the Kleinian sigma function are called the generalized Weierstrass functions

$$\wp_{ij}(z) = -\frac{\partial}{\partial z_i}\frac{\partial}{\partial z_j} \log \sigma(z) = \frac{\sigma_i(z)\sigma_j(z) - \sigma(z)\sigma_{ij}(z)}{\sigma^2(z)}, \qquad (2.4.6)$$

where $\sigma_i(z)$ denotes the derivative of the Kleinian sigma function with respect to the i-th component of z.

The solution of Jacobi's inversion problem (2.4.3) can be given in terms of generalized Weierstrass functions. Let X be the Riemann surface of $\sqrt{P_n}$. By a rational transformation it can be achieved that $P_n(x) = \sum_{j=0}^{2g+1} b_j x^j$, where g is the genus of X, i.e. $g = \left[\frac{n-1}{2}\right]$. Then the components of the solution $x = (x_1, \ldots, x_g)^t \in S^g(X)$ of Jacobi's inversion problem (2.4.3) are given by the g solutions of [48]

$$\frac{b_{2g+1}}{4}x^g - \sum_{i=1}^{g} \wp_{gi}(y)x^{i-1} = 0, \qquad (2.4.7)$$

where $y \in \mathrm{Jac}(X)$ is the left hand side of (2.4.3). In the case of $g = 2$, we can rewrite this result with the help of the theorems of Vieta in the form

$$\begin{aligned} x_1 + x_2 &= \frac{4}{b_5}\wp_{22}(y), \\ x_1 x_2 &= -\frac{4}{b_5}\wp_{12}(y). \end{aligned} \qquad (2.4.8)$$

This is the starting point for finding analytical solutions of geodesic equations in space-times with a nonvanishing cosmological constant. The idea is to consider the geodesic equation $s - s_0 = \int_{x_0}^{x} dz_2$ as a part of Jacobi's inversion problem

$$\begin{aligned} y_1 &= \int_{x_0}^{x} dz_1 + \int_{x_0}^{x_2} dz_1, \\ s - s_0 = y_2 &= \int_{x_0}^{x} dz_2 + \int_{x_0}^{x_2} dz_2, \end{aligned} \qquad (2.4.9)$$

but reducing these equations by a limiting process $x_2 \to x_0$. It can then be shown that $\int_{x_0}^{x_1} dz$ is an element of the one-dimensional theta divisor $\Theta_{K_{x_0}}$, which allows to express $\int_{x_0}^{x_1} dz_1$ as a function of $\int_{x_0}^{x_1} dz_2$ and vice versa. Together with Eqs. (2.4.8) and (2.4.6) the analytical solution of the geodesic equation can then be expressed in terms of the derivatives of the Kleinian sigma function. The details

2.4. Solution methods

of this solution method are explained in Sec. 3.3.1.2. Some mathematical background can be found in [52].

However, also the more complicated equations of motion of type (2.2.11) have to be solved. In Eq. (2.2.13) it was shown that this type of equation can be reformulated in terms of the canonical differential of third kind $dP(x_1, x_2)$. A particular useful feature of $dP(x_1, x_2)$ is that it can be expressed in terms of Kleinian sigma functions.

This can be shown in two steps. First, the differential of first kind $dP(x_1, x_2)$ can be rewritten in terms of the holomorphic and meromorphic differentials as well as the fundamental 2-differential defined by

$$d\hat{\omega}(x_1, x_2) = \frac{2y_1 y_2 + F(x_1, x_2)}{4(x_1 - x_2)^2} \frac{dx_1}{y_1} \frac{dx_2}{y_2}, \quad (2.4.10)$$

for $y_j^2 = P_n(x_j) = \sum_{i=0}^{2g+2} \lambda_i x_j^i$ and

$$F(x_1, x_2) = 2\lambda_{2g+2} x_1^{g+1} x_2^{g+1} + \sum_{i=0}^{g} x_1^i x_2^i (2\lambda_{2i} + \lambda_{2i+1}(x_1 + x_2)), \quad (2.4.11)$$

see [50, 48]. The connection

$$\int_{u_0}^{u} dP(z, c) = \int_{u_0}^{u} \int_{c}^{z} d\hat{\omega}(u, z_1) - \left(\int_{u_0}^{u} dz\right)^t \left(\int_{c}^{z} dr\right)$$
$$= \frac{1}{2} \sum_{i=1}^{2} \int_{u_0}^{u} \int_{c_i}^{z_i} d\hat{\omega}(x, z_i) - \left(\int_{u_0}^{u} dz\right)^t \left(\int_{c}^{z} dr\right) \quad (2.4.12)$$

can be proved by differentiation by u and z, but see also [50]. In turn, the fundamental 2-differential $d\hat{\omega}$ can be related to the quotient of θ functions

$$x \mapsto \frac{\theta\left(\int_{x_0}^{x} dv - (\sum_{i=1}^{2} \int_{x_0}^{z_i} dv - K_{x_0})\right)}{\theta\left(\int_{x_0}^{x} dv - (\sum_{i=1}^{2} \int_{x_0}^{c_i} dv - K_{x_0})\right)}. \quad (2.4.13)$$

which has simple zeros in z_i and simple poles in c_i by Riemann's vanishing theorem, see (2.3.6) or [50]. The same zeros and poles can be found for the function

$$x \mapsto \exp\left[\int_{x_0}^{x} \int_{c}^{z} (d\hat{\omega}(x, z) + 2dz(x)\eta(2\omega)^{-1} dz(z))\right], \quad (2.4.14)$$

what implies that they are equal up to a constant. Then the quotient of theta functions (2.4.13) can be rewritten in terms of the Kleinian sigma function. The final result is [50, 48]

$$\int_{\infty}^{x} \left(\sum_{i=1}^{g} \int_{c_i}^{z_i} d\hat{\omega}(x, z_i)\right)$$
$$= \log\left\{\frac{\sigma\left(\int_{\infty}^{x} dz - \sum_{i=1}^{g} \int_{a_i}^{z_i} dz\right)}{\sigma\left(\int_{\infty}^{x} dz - \sum_{i=1}^{g} \int_{a_i}^{c_i} dz\right)}\right\} - \log\left\{\frac{\sigma\left(-\sum_{i=1}^{g} \int_{a_i}^{z_i} dz\right)}{\sigma\left(-\sum_{i=1}^{g} \int_{a_i}^{c_i} dz\right)}\right\}. \quad (2.4.15)$$

2. Mathematical preliminaries

The Eqs. (2.4.12) and (2.4.15) can be used to express geodesic equations of type (2.2.11) in terms of integrals of first and second kind as well as by logarithms of the Kleinian sigma function. This can be seen as an analogue to the solution method for elliptic integrals of type (2.1.17) presented in Sec. 2.1.2.

CHAPTER 3

Geodesics in spherically symmetric space-times

In this chapter we will study geodesic motion in static and spherically symmetric vacuum solutions of the Einstein field equation

$$R_{\mu\nu} - \frac{1}{2}R g_{\mu\nu} + \Lambda g_{\mu\nu} = \kappa T_{\mu\nu} \tag{3.0.1}$$

with and without cosmological constant Λ given by a metric

$$ds^2 = \frac{\Delta}{r^2}dt^2 - \frac{r^2}{\Delta}dr^2 - r^2(d\theta^2 + \sin^2\theta d\varphi^2), \tag{3.0.2}$$

where $\Delta = r^2 - 2Mr - \frac{1}{3}\Lambda r^4 + Q^2$ depends on the radial coordinate r only. (Here and in the following sections units are used where the speed of light in vacuum c and the gravitational constant G are equal to 1.) This space-time is characterized by the mass M, the cosmological constant Λ, and the electric charge Q of the black hole. All space-times with metrics of type (3.0.2) have a singularity located at $r = 0$, which could be a naked singularity for some choices of Λ and Q. This case is widely neglected except for a few remarks concerning null geodesics in space-times with $\Lambda = 0, Q \neq 0$, but see also [53].

The geodesic motion in such a space-time is described by the geodesic equation

$$0 = \frac{d^2 x^\mu}{ds^2} + \{^{\mu}_{\rho\sigma}\} \frac{dx^\rho}{ds}\frac{dx^\sigma}{ds}, \tag{3.0.3}$$

where

$$\{^{\mu}_{\rho\sigma}\} = \frac{1}{2}g^{\mu\nu}\left(\partial_\rho g_{\sigma\nu} + \partial_\sigma g_{\rho\nu} - \partial_\nu g_{\rho\sigma}\right) \tag{3.0.4}$$

3. Geodesics in spherically symmetric space-times

is the Christoffel symbol. Although this chapter focuses on geodesic motion in static and spherically symmetric space-times with a nonvanishing cosmological constant, geodesics in space-times with zero cosmological constant are studied to permit a direct comparison between these two cases. For a more detailed discussion of geodesic motion in space-times with a vanishing cosmological constant see for example [6, 14, 15].

Before analytically solving Eq. (3.0.3) a method for classifying the different geometric types of geodesics in the various subclasses of (3.0.2) will be presented in Sec. 3.1. Also, some global properties of geodesics for all subclasses of (3.0.2) are considered.

The geodesic equation for the most simple spherically symmetric space-time, the Schwarzschild space-time, given by (3.0.2) with $\Delta = \Delta_{\rm S} := r^2 - 2Mr$ [2],

$$ds^2 = \left(1 - \frac{2M}{r}\right) dt^2 - \left(1 - \frac{2M}{r}\right)^{-1} dr^2 - r^2(d\theta^2 + \sin^2\theta d\varphi^2), \qquad (3.0.5)$$

will be discussed in Sec. 3.2. In 1931, Hagihara was the first who solved the geodesic equations in these space-time analytically [6]. The method he used for this solution is based on the theory of elliptic functions presented in the Chap. 2. In this book the geodesic equations in Schwarzschild space-time will be solved analogously to his method. The metric (3.0.5) can be generalized to include black holes which possess an electric charge Q. The geodesic equations in these Reissner-Nordström space-times given by (3.0.2) with $\Delta = \Delta_{\rm RN} := r^2 - 2Mr + Q^2$ [7, 8],

$$ds^2 = \left(1 - \frac{2M}{r} + \frac{Q}{r^2}\right) dt^2 - \left(1 - \frac{2M}{r} + \frac{Q}{r^2}\right)^{-1} dr^2 - r^2(d\theta^2 + \sin^2\theta d\varphi^2), \qquad (3.0.6)$$

possess the same mathematical structure as geodesics in Schwarzschild space-time and, thus, can be solved analogously. Furthermore, properties and types of geodesics special for Schwarzschild and Reissner-Nordström space-times will be discussed in this section. All possible orbits in Schwarzschild and Reissner-Nordström space-time are classified in terms of the energy and the angular momentum of the test particle or light ray and, in the case of $Q \neq 0$, dependent on the electric charge of the black hole.

The structure of geodesic equations in spherically symmetric black hole space-times with a non-vanishing cosmological constant Λ is much more complicated and, thus, they cannot be solved with the method used in [6]. Indeed, Sec. 3.3 focuses on the analytical solution of the geodesic equations in Kottler space-times (also known as Schwarzschild-(anti-)de Sitter space-times) given by (3.0.2) with $\Delta = \Delta_{\rm SdS} := r^2 - 2Mr - \frac{1}{3}\Lambda r^4$ [54, 55],

$$ds^2 = \left(1 - \frac{2M}{r} - \frac{1}{3}\Lambda r^2\right) dt^2 - \left(1 - \frac{2M}{r} - \frac{1}{3}\Lambda r^2\right)^{-1} dr^2 - r^2(d\theta^2 + \sin^2\theta d\varphi^2), \qquad (3.0.7)$$

in terms of the more general theory of hyperelliptic functions outlined in Chap. 2. The resulting new solution is then used to derive an analytical expression of the periastron advance of bound orbits together with a post-Schwarzschild expansion for small Λ, and to address the interesting question whether the Pioneer Anomaly [26] is connected to a cosmological force. Furthermore, the analytical solution of the geodesic equation in Reissner-Nordström-de Sitter space-times given by (3.0.2) with $\Delta = \Delta_{\text{RNdS}} := r^2 - 2Mr - \frac{1}{3}\Lambda r^4 + Q^2$ [12],

$$ds^2 = \left(1 - \frac{2M}{r} - \frac{1}{3}\Lambda r^2 + \frac{Q^2}{r^2}\right)dt^2 - \left(1 - \frac{2M}{r} - \frac{1}{3}\Lambda r^2 + \frac{Q^2}{r^2}\right)^{-1}dr^2$$
$$- r^2(d\theta^2 + \sin^2\theta d\varphi^2), \quad (3.0.8)$$

is presented. For each of these space-times possible orbit types are classified in terms of energy and angular momentum and dependent on the parameters of the black hole. This classification is compared to the case of a vanishing cosmological constant and its influence on the orbit types whithin the different classes of geodesic motion is discussed.

Finally, it will be demonstrated that the methods developed and applied in this chapter can also be used to discuss geodesics in higher-dimensional spherically symmetric and static space-times. As an example, geodesics in six-dimensional Schwarzschild space-time [56] are classified and the analytical solutions of the geodesic equations are presented. Even more general higher-dimensional space-times are treated in [53].

3.1 General types of geodesics

For the geodesic equation (3.0.3) in the general space-time described by (3.0.2) three constants of motions can be identified. A fourth constant is not necessary as (3.0.3) can be restricted to the equatorial plane because of the spherical symmetry. The first constant of motion is given by the normalization condition $g_{\mu\nu}\frac{dx^\mu}{ds}\frac{dx^\nu}{ds} = \epsilon$ where for massive particles $\epsilon = 1$ and for light $\epsilon = 0$. Furthermore, the existence of the Killing vectors $\frac{\partial}{\partial \varphi}$ and $\frac{\partial}{\partial t}$ following from the fact that the metric (3.0.2) is static and spherically symmetric induces the conserved energy and angular momentum

$$E = g_{tt}\frac{dt}{ds} = \frac{\Delta}{r^2}\frac{dt}{ds},$$
$$L = r^2\frac{d\varphi}{ds}, \quad (3.1.1)$$

which reduce the geodesic equation to one ordinary differential equation

$$\left(\frac{dr}{ds}\right)^2 = E^2 - \frac{\Delta}{r^2}\left(\epsilon + \frac{L^2}{r^2}\right). \quad (3.1.2)$$

Note that here all test particles are assumed to have no electric or magnetic charge and, thus, to describe a geodesic in the considered space-time. However, this case may be treated analogously as the equations of motions are not significantly changed, see [14, 57].

3. Geodesics in spherically symmetric space-times

Together with energy and angular momentum conservation we obtain the corresponding equations for r as functions of φ and t

$$\left(\frac{dr}{d\varphi}\right)^2 = \frac{r^4}{L^2}\left(E^2 - \frac{\Delta}{r^2}\left(\epsilon + \frac{L^2}{r^2}\right)\right) =: R(r), \qquad (3.1.3)$$

$$\left(\frac{dr}{dt}\right)^2 = \frac{1}{E^2}\frac{\Delta^2}{r^4}\left(E^2 - \frac{\Delta}{r^2}\left(\epsilon + \frac{L^2}{r^2}\right)\right). \qquad (3.1.4)$$

Eqs. (3.1.2)-(3.1.4) give a complete description of the dynamics of the geodesic motion. Eq. (3.1.2) suggests the introduction of an effective potential

$$\left(\frac{dr}{ds}\right)^2 = E^2 - V_{\text{eff}}, \quad \text{with} \quad V_{\text{eff}} = \frac{\Delta}{r^2}\left(\epsilon + \frac{L^2}{r^2}\right). \qquad (3.1.5)$$

The shape of an orbit depends on the energy E and the angular momentum L of the test particle or light ray under consideration as well as the cosmological constant Λ and the electric charge Q (the mass can be absorbed through a rescaling of the radial coordinate). These quantities are all contained in the polynomial $R(r)$ defined in Eq. (3.1.3). Since r should be real and positive the physically acceptable regions are given by those r for which $E^2 \geq V_{\text{eff}}$ owing to the square on the left hand side of (3.1.2). Hence, the number of positive real zeros of R uniquely characterizes the form of the resulting orbits.

The following different types of orbits can be identified in space-times described by the metric (3.0.2), see also Fig. 3.1.

(i) Flyby orbits: r starts from ∞, then approaches a periapsis $r = r_p$ and goes back to ∞.

(ii) Bound orbits: r oscillates between to boundary values $r_p \leq r \leq r_a$ with $0 < r_p < r_a < \infty$.

(iii) Terminating bound orbits: r starts in $(0, r_a]$ for $0 < r_a < \infty$ and falls into the singularity at $r = 0$.

(iv) Terminating escape orbits: r comes from ∞ and falls into the singularity at $r = 0$.

All other types of orbits are exceptional and treated separately. They are connected with the appearance of multiple zeros in R or with parameter values which reduce the degree of R. In both cases the structure of the differential equation (3.1.3) is considerably simplified. These orbits are radial geodesics with $L = 0$ (i.e. $\frac{dr}{d\varphi} = 0$), circular orbits with constant r, orbits asymptotically approaching circular orbits, and in the case of $\Lambda = 0$ parabolic orbits with $E^2 = 1$.

The four regular types of geodesic motion correspond to different arrangements of the real and positive zeros of R defining the borders of $R(r) \geq 0$ or, equivalently, $E^2 \geq V_{\text{eff}}$. If $R(r)$ has no real

3.1. General types of geodesics

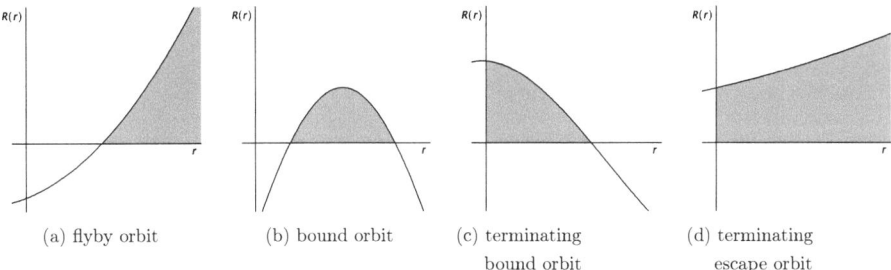

Figure 3.1: Possible arrangements of real positive zeros of $R(r)$ in spherical symmetric space-times. The allowed regions of geodesic motion are shaded in gray. The zeros correspond to $V_{\text{eff}} = E^2$. Combinations of (a) - (c) are possible, e.g. simultaneously a bound and a terminating bound orbit.

and positive zeros at all a terminating escape orbit is possible if $R(r) > 0$ for all $r > 0$, but else no geodesic motion is allowed. If $R(r)$ has at least one real and positive zero then a flyby orbit is possible if $\lim_{r \to \infty} R(r) = \infty$, and a terminating bound orbit if $R(r) > 0$ for $0 < r < r_1$ where r_1 is the smallest positive zero. If $R(r)$ has at least two real zeros $r_1 < r_2$ with $R(r) > 0$ for $r_1 < r < r_2$ a bound orbit is permitted. In Fig. 3.1 different arrangements of real zeros of R and the resulting types of orbits are shown. If R is such that multiple types of orbits are possible the actual orbit depends on the initial position of the test particle or light ray.

For a fixed parameter combination either a terminating escape orbit or any of the other orbit types is allowed. In Schwarzschild and Schwarzschild-de Sitter space-times all 4 types of orbits can be realized for certain parameter combinations. However, in Reissner-Nordström and Reissner-Nordström-de Sitter space-times no terminating orbits are possible. The reason is the electric charge, which leads to gravitational repulsion near the singularity at $r = 0$. Accordingly, we have either flyby or bound orbits. The only exception to this rule are the radial null geodesics, where $r = \pm Es + \text{constant}$ and, thus, a co-moving observer will arrive at the singularity at a finite time. (However, a distant observer will never witness a fall through the event horizon as there the coordinate time t increases to infinity, cf. [14].) A particular feature is that the Reissner-Nordström(-de Sitter) space-times are the only one considered in this chapter which allow 2 different bound orbits. In Reissner-Nordström space-times both bound orbits are characterized by the same period which is a consequence of the fact that the solution, owing to the order of the polynomial, is given in terms of the same Weierstrass $\wp$ function.

3. Geodesics in spherically symmetric space-times

3.2 Schwarzschild and Reissner-Nordström space-times

This section deals with the geodesic equation (3.0.3)

$$0 = \frac{d^2 x^\mu}{ds^2} + \{^{\mu}_{\rho\sigma}\} \frac{dx^\rho}{ds} \frac{dx^\sigma}{ds}$$

in Schwarzschild and Reissner-Nordström space-times given by the metrics (3.0.5) and (3.0.6), respectively, and characterized by the mass M and, in the case of Reissner-Nordström space-time, by the electric charge Q of the black hole. With the normalization condition $g_{\mu\nu} \frac{dx^\mu}{ds} \frac{dx^\nu}{ds} = \epsilon$ where for massive particles $\epsilon = 1$ and for light $\epsilon = 0$, with conserved energy and angular momentum (3.1.1), and the restriction to the equatorial plane the geodesic equation reduces to (3.1.2)-(3.1.4)

$$\left(\frac{dr}{d\varphi}\right)^2 = R(r) = \frac{r^4}{L^2}\left(E^2 - \frac{\Delta}{r^2}\left(\epsilon + \frac{L^2}{r^2}\right)\right), \tag{3.1.3}$$

$$\left(\frac{dr}{ds}\right)^2 = E^2 - \frac{\Delta}{r^2}\left(\epsilon + \frac{L^2}{r^2}\right), \tag{3.1.2}$$

$$\left(\frac{dr}{dt}\right)^2 = \frac{1}{E^2}\frac{\Delta^2}{r^4}\left(E^2 - \frac{\Delta}{r^2}\left(\epsilon + \frac{L^2}{r^2}\right)\right). \tag{3.1.4}$$

Here $R(r)$ is a polynomial of degree 3 for Schwarzschild space-time and of degree 4 for Reissner-Norström space-times. Therefore, in both cases Eq. (3.1.3) can be solved in terms of elliptic functions. In the following we will treat Eq. (3.1.3) for Schwarzschild and Reissner-Nordström space-times separately.

3.2.1 Geodesics in Schwarzschild space-times

3.2.1.1 Types of orbits

As explained in Sec. 3.1, all possible types of orbits in Schwarzschild space-time can be determined from the right hand side of Eq. (3.1.3)

$$\left(\frac{dr}{d\varphi}\right)^2 = R(r) = \frac{r^4}{L^2}\left(E^2 - \frac{\Delta_S}{r^2}\left(\epsilon + \frac{L^2}{r^2}\right)\right), \tag{3.1.3}$$

where $\Delta_S = r^2 - 2Mr$. For the analysis of the dependence of the possible types of orbits on the parameters of the space-time and the test particle or light ray it is convenient to use dimensionless quantities. Thus, we introduce

$$\bar{r} := \frac{r}{M}, \quad \mathcal{L} := \frac{M^2}{L^2}, \tag{3.2.1}$$

and rewrite Eq. (3.1.3) as

$$\left(\frac{d\bar{r}}{d\varphi}\right)^2 = (E^2 - \epsilon)\mathcal{L}\bar{r}^4 + 2\epsilon\mathcal{L}\bar{r}^3 - \bar{r}^2 + 2\bar{r} =: R_S(\bar{r}). \tag{3.2.2}$$

3.2. Schwarzschild and Reissner-Nordström space-times

In the following we will analyse possible types of orbits dependent on the parameters of the test particle or light ray ϵ, E^2, and $\mathcal{L}$. The major point in this analysis is that (3.2.2) implies $R_S(\bar{r}) \geq 0$ as a necessary condition for the existence of a geodesic. Thus, the zeros of R_S are extremal values of $\bar{r}(\varphi)$ and determine (together with the sign of R_S between two zeros) the type of geodesic. The polynomial R_S is in general of degree 4 and, therefore, has 4 (complex) zeros of which the positive real zeros are of interest for the type of orbit. As $\bar{r} = 0$ is a zero of R_S for all values of the parameters, this zero is neglected in the following and

$$R_S^*(\bar{r}) := (E^2 - \epsilon)\mathcal{L}\bar{r}^3 + 2\epsilon\mathcal{L}\bar{r}^2 - \bar{r} + 2 \tag{3.2.3}$$

is considered instead of R_S.

All types of orbit in Schwarzschild space-time were extensively discussed by Hagihara [6]. Nevertheless, let us examine the different possible orbits so that they can be directly compared with orbits in the other space-times considered in this chapter.

Dependent on the number of positive real zeros and the sign of $(E^2 - \epsilon)$ the following types of orbits are possible

(a) $(E^2 - \epsilon) > 0$, i.e. $\lim_{\bar{r} \to \infty} R_S^*(\bar{r}) = \infty$

   (i) 0 positive real zeros: terminating escape orbit,

   (ii) 1 positive real zero: flyby orbit,

   (iii) 2 positive real zeros: flyby and terminating bound orbit,

   (iv) 3 positive real zeros: flyby and bound orbit.

(b) $(E^2 - \epsilon) < 0$, i.e. $\lim_{\bar{r} \to \infty} R_S^*(\bar{r}) = -\infty$

   (i) 0 positive real zeros: no geodesic motion possible,

   (ii) 1 positive real zero: terminating bound orbit,

   (iii) 2 positive real zeros: bound orbit,

   (iv) 3 positive real zeros: bound and terminating bound orbit.

If more than one orbit type is possible the actual orbit depends on the initial position of the test particle or light ray.

For a given set of parameters ϵ, E^2, and $\mathcal{L}$ the polynomial R_S^* has a certain number of positive real zeros. If E^2 and $\mathcal{L}$ are varied this number can change only if two zeros of R_S^* merge to one. (A positive

3. Geodesics in spherically symmetric space-times

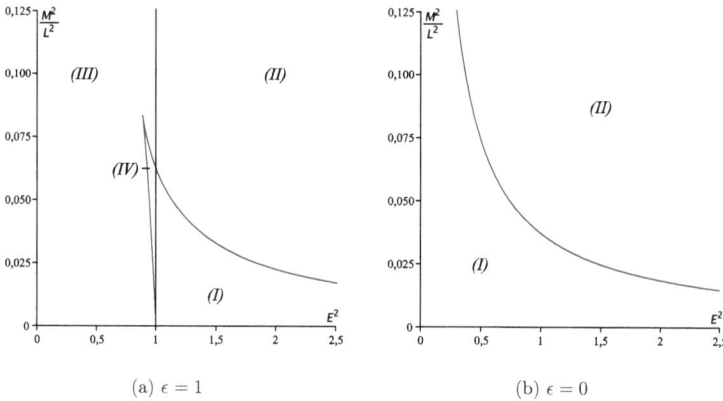

Figure 3.2: Regions of different types of geodesic motion in Schwarzschild space-time for test particles ($\epsilon = 1$) and light ($\epsilon = 0$). For effective potentials see Fig. 3.3.

zero can not become negative because R_S^* is a continuous function of E^2 or $\mathcal{L}$ and $R_S^*(0) = 2$). This merger happens at $\bar{r} = x$ iff

$$R_S^* = (\bar{r} - x)^2 (a_1 \bar{r} + a_0) \tag{3.2.4}$$

for some real constants a_0, a_1. By a comparison of coefficients we can solve the resulting 4 equations for E^2 and $\mathcal{L}$ dependent on ϵ. For $\epsilon = 1$ this yields

$$E^2(x) = \frac{(x-2)^2}{x(x-3)}, \quad \mathcal{L}(x) = \frac{x-3}{x^2}, \tag{3.2.5}$$

where x is the position of the double zero, and for $\epsilon = 0$

$$\mathcal{L} = \frac{1}{27 E^2}. \tag{3.2.6}$$

(Note that this procedure is equivalent to solving $D = 0$ for E and $\mathcal{L}$, where D is the discriminant of R_S^*.)

In Fig. 3.2 the results of this analysis are shown for both test particles ($\epsilon = 1$) and light ($\epsilon = 0$). Here we can identify 4 regions of different types of geodesic motion:

(I) $R_S^*(\bar{r})$ has 2 positive real zeros $r_1 < r_2$ with $R_S(\bar{r}) > 0$ for $0 \leq \bar{r} \leq r_1$ and $r_2 \leq \bar{r}$. Possible orbit types: flyby and terminating bound orbits.

(II) $R_S^*(\bar{r})$ has 0 positive real zeros and $R_S(\bar{r}) \geq 0$ for $0 \leq \bar{r}$. Possible orbit types: terminating escape orbits.

3.2. Schwarzschild and Reissner-Nordström space-times

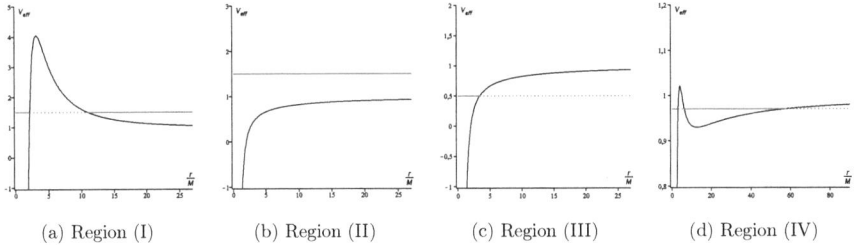

(a) Region (I)   (b) Region (II)   (c) Region (III)   (d) Region (IV)

Figure 3.3: Effective potentials for different regions of geodesic motion in Schwarzschild space-time. The horizontal lines denote the squared energy parameter E^2.

(III) $R_S^*(\bar{r})$ has 1 positive real zero r_1 with $R_S(\bar{r}) \geq 0$ for positive r. Possible orbit types: terminating bound orbits.

(IV) $R_S^*(\bar{r})$ has 3 positive real zeros $r_1 < r_2 < r_3$ with $R_S(\bar{r}) \geq 0$ for $0 \leq \bar{r} \leq r_1$ and $r_2 \leq \bar{r} \leq r_3$. Possible orbit types: bound and terminating bound orbits.

For light rays only regions (I) and (II) appear. In the case of test particles $\epsilon = 1$ the straight line $E^2 = 1$ divides the plot in two parts. For $E^2 < 1$ the polynomial R_S^* tends to $-\infty$ for $r \to \infty$, i.e. we may not have any escape orbits in these regions. However, for $E^2 > 1$ it is $\lim_{r \to \infty} R_S^*(r) = \infty$ and, thus, there is always an orbit which reaches infinity. For every region, examples of effective potentials are displayed in Fig. 3.3, of timelike geodesics in Fig. 3.4, and of null geodesics in Fig. 3.5. Each of the orbits was plotted using the analytical solution of $r(\varphi)$ derived in the next subsection. A summary of possible orbit types can be found in Tab. 3.1

For light as well as test particles exceptional orbits appear at the boundaries of the regions (I) to (IV) corresponding to multiple zeros of R_S^* or to $E^2 = 1$. In the case of multiple zeros the boundary is described by E^2 and $\mathcal{L}$ given by (3.2.5) or (3.2.6), respectively, and the corresponding test particle or light ray moves on a circular orbit, which may be stable or unstable. For test particles, the substitution of Eq. (3.2.5) in $\frac{d^2}{d\bar{r}^2} R_S^*(\bar{r})$ yields

$$\frac{d^2}{dx^2} R_S^*(x) = -\frac{2(x-6)}{x^2}, \qquad (3.2.7)$$

i.e. the double zero x of R_S^* is a minimum if $x < 6$ whereas it is a maximum if $x > 6$. The parameters E^2 and $\mathcal{L}$ given by a double zero x which is a minimum correspond to unstable circular orbits, where an asymptotical approach is possible for $\bar{r} < x$ and $\bar{r} > x$. Parameters E^2 and $\mathcal{L}$ given by a double zero which is a maximum correspond to stable circular orbits, where an asymptotic approach is not possible. The triple zero $x = 6$ (or $r = 6M$) is a saddle point and corresponds to the corner at $E^2 = \frac{8}{9}$ and $\mathcal{L} = \frac{1}{12}$ in the boundary of region (IV). This is the innermost stable circular orbit. For light rays

3. Geodesics in spherically symmetric space-times

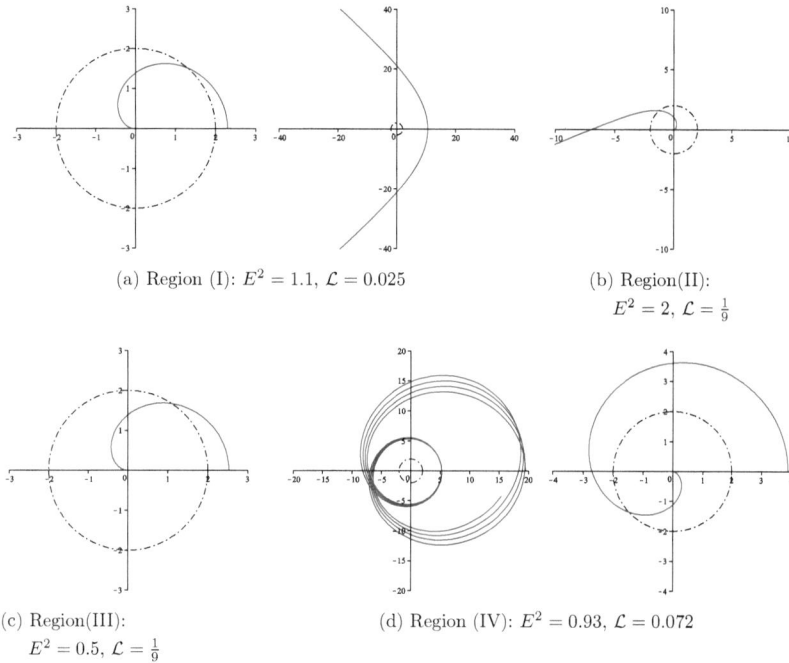

Figure 3.4: Timelike geodesic in Schwarzschild space-time for every region of orbit types. The plot is in units of M and dashed lines mark the horizon.

Subcaptions:
(a) Region (I): $E^2 = 1.1$, $\mathcal{L} = 0.025$
(b) Region(II): $E^2 = 2$, $\mathcal{L} = \frac{1}{9}$
(c) Region(III): $E^2 = 0.5$, $\mathcal{L} = \frac{1}{9}$
(d) Region (IV): $E^2 = 0.93$, $\mathcal{L} = 0.072$

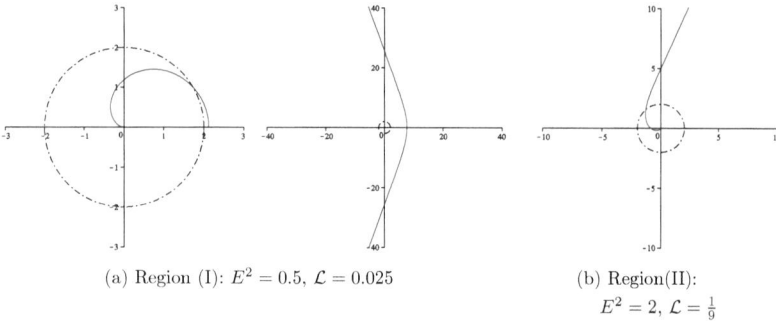

(a) Region (I): $E^2 = 0.5$, $\mathcal{L} = 0.025$
(b) Region(II): $E^2 = 2$, $\mathcal{L} = \frac{1}{9}$

Figure 3.5: Null geodesics in Schwarzschild space-time for every region of orbit types. The plot is in units of M and dashed lines mark the horizon.

3.2. Schwarzschild and Reissner-Nordström space-times

region	pos. zeros	range of $\bar{r}$	types of orbits
I	2		flyby, terminating bound
II	0		terminating escape
III	1		terminating bound
IV	3		bound, terminating bound

Table 3.1: Orbit types in Schwarzschild space-time. The second column gives the number of positive zeros of the polynomial $R_S^*(\bar{r})$. In the third column, the thick lines represent the range of orbits and turning points are shown by thick dots. The small vertical line denotes $\bar{r} = 0$.

double zeros can only be located at $x = 3$ (or $r = 3M$), which is always a minimum and, thus, corresponds to an unstable circular orbit. In the case of parabolic orbits with $E^2 = \epsilon = 1$ the degree of R_S^* is reduced to 2 and R_S^* has 2 real and positive zeros if $\mathcal{L} < \frac{1}{16}$, a double zero at $x = 4$ if $\mathcal{L} = \frac{1}{16}$, and no real zeros at all if $\mathcal{L} > \frac{1}{16}$. Thus, for $\mathcal{L} < \frac{1}{16}$ a flyby and a terminating bound orbit is possible, for $\mathcal{L} = \frac{1}{16}$ there is an unstable circular orbit at $r = 4M$, and for $\mathcal{L} > \frac{1}{16}$ only a terminating escape orbit is allowed.

3.2.1.2 Analytical solution of geodesic equations

Now the analytical solution of the geodesic equation (3.1.3) is presented. As can be seen from Eq. (3.2.2),

$$\left(\frac{d\bar{r}}{d\varphi}\right)^2 = R_S(\bar{r}) = (E^2 - \epsilon)\mathcal{L}\bar{r}^4 + 2\epsilon\mathcal{L}\bar{r}^3 - \bar{r}^2 + 2\bar{r}, \tag{3.2.2}$$

the right hand side of this equation is in general a polynomial of degree 4 and, therefore, the differential equation is of elliptic type if R_S has only simple zeros but can be solved in terms of elementary functions if R_S has multiple zeros. In the latter case the analytical solution can be found for example in [14, 31]. In the following it is assumed that R_S has only simple zeros.

As usual, we introduce a new variable $u = M/r$ (which is also the standard substitution for problems of type (2.1.9)) and obtain from Eq. (3.2.2)

$$\left(\frac{du}{d\varphi}\right)^2 = 2u^3 - u^2 + 2\epsilon\mathcal{L}u + \mathcal{L}(E^2 - \epsilon) \tag{3.2.8}$$

with the dimensionless parameter $\mathcal{L} := \frac{M^2}{L^2}$ introduced in Eq. (3.2.1).

With the standard substitution $u = 2y + \frac{1}{6}$ Eq. (3.2.8) can be transformed to the Weierstrass form

3. Geodesics in spherically symmetric space-times

(2.1.5)

$$\left(\frac{dy}{d\varphi}\right)^2 = 4y^3 - g_2 y - g_3 \,, \tag{3.2.9}$$

where

$$\begin{aligned} g_2 &= \frac{1}{12} - \epsilon \mathcal{L} \,, \\ g_3 &= \frac{1}{216} - \frac{1}{12}\epsilon\mathcal{L} - \frac{1}{4}\mathcal{L}(E^2 - \epsilon) \,. \end{aligned} \tag{3.2.10}$$

Then the analytical solution of Eq. (3.1.3) for Schwarzschild space-time is given by

$$r(\varphi) = \frac{M}{2y(\varphi) + \frac{1}{6}} = \frac{M}{2\wp(\varphi - \varphi_{in}) + \frac{1}{6}} \,, \tag{3.2.11}$$

where

$$\varphi_{in} = \varphi_0 + \int_{y_0}^{\infty} \frac{dz}{\sqrt{4z^3 - g_2 z - g_3}} \,, \quad y_0 = \frac{1}{2}\left(\frac{M}{r_0} - \frac{1}{6}\right) \,, \tag{3.2.12}$$

depends only on the initial values φ_0 and r_0. In Figs. 3.4 and 3.5 this solution was used to create the examples of timelike and null geodesics for each region of different types of orbits identified in the foregoing subsection.

3.2.2 Geodesics in Reissner-Nordström space-times

3.2.2.1 Types of orbits

In this subsection possible types of geodesics in Reissner-Nordström space-time are considered. They can be determined from the right hand side of Eq. (3.1.3),

$$\left(\frac{dr}{d\varphi}\right)^2 = R(r) = \frac{r^4}{L^2}\left(E^2 - \frac{\Delta_{\text{RN}}}{r^2}\left(\epsilon + \frac{L^2}{r^2}\right)\right) \,, \tag{3.1.3}$$

where $\Delta_{\text{RN}} = r^2 - 2Mr + Q^2$, as explained in Sec. 3.1. (Remember that test particles are assumed to be neutral.) Just like in Schwarzschild space-time, it is convenient to use dimensionless quantities

$$\bar{r} := \frac{r}{M} \,, \quad \mathcal{L} := \frac{M^2}{L^2} \,, \quad \bar{Q} = \frac{Q}{M} \,. \tag{3.2.13}$$

for the analysis. In terms of these quantities Eq. (3.1.3) for Reissner-Nordström space-time reads

$$\left(\frac{d\bar{r}}{d\varphi}\right)^2 = (E^2 - \epsilon)\mathcal{L}\bar{r}^4 + 2\epsilon\mathcal{L}\bar{r}^3 - (\epsilon\mathcal{L}\bar{Q}^2 + 1)\bar{r}^2 + 2\bar{r} - \bar{Q}^2 =: R_{\text{RN}}(\bar{r}) \,. \tag{3.2.14}$$

In Reissner-Nordström space-time there are two horizons located at the two zeros $r_\pm$ of Δ_{RN}, $r_\pm = M \pm \sqrt{M^2 - Q^2}$. These horizons are real if $Q^2 \leq M^2$ (with $r_+ = r_-$ for $Q^2 = M^2$) or,

3.2. Schwarzschild and Reissner-Nordström space-times

equivalently, $\bar{Q}^2 \leq 1$. For other choices of $\bar{Q}^2$ no screen prevents an observer to see the naked singularity at $r = 0$. In the following we will assume that $\bar{Q}^2 \leq 1$ if not stated otherwise.

As already explained in Sec. 3.1, the real and positive zeros of R_{RN} are turning points of the geodesic motion and, thus, determine the type of orbit. The number of real and positive zeros of R_{RN} changes for varying parameters E^2, $\mathcal{L}$, and $\bar{Q}$ if two zeros merge to one, i.e. at the parameters for which R_{RN} has double zeros. (Again, a positive real zero may not become negative with the same argument as in Sec. 3.2.1.1.) A double zero x of R_{RN} fulfills the equation

$$R_{\mathrm{RN}} = (\bar{r} - x)^2 (a_1 \bar{r} + a_0), \qquad (3.2.15)$$

which can be solved by a comparison of coefficients for E^2 and $\mathcal{L}$ dependent on the parameters ϵ and $\bar{Q}^2$ of the black hole. For $\epsilon = 1$ this yields

$$E^2(x) = \frac{(x(x-2) + \bar{Q}^2)^2}{x^2(x^2 - 3x + 2\bar{Q}^2)}, \quad \mathcal{L}(x) = \frac{x^2 - 3x + 2\bar{Q}^2}{x^2(x - \bar{Q}^2)}, \qquad (3.2.16)$$

where x is the position of the double zero, and for $\epsilon = 0$

$$\mathcal{L} = \frac{2(1 + \sqrt{9 - 8\bar{Q}^2})}{E^2 (3 + \sqrt{9 - 8\bar{Q}^2})^3}. \qquad (3.2.17)$$

From Eq. (3.2.17) it is obvious that $\mathcal{L}$ is imaginary for $\bar{Q}^2 > \frac{9}{8}$ and, therefore, that double zeros are not possible for $\bar{Q}^2 > \frac{9}{8}$ (or $Q^2 > \frac{9M^2}{8}$) and $\epsilon = 0$.

In Fig. 3.6 regions of different types of geodesic motion in Reissner-Nordström space-time are shown for varying $\bar{Q}^2$ and ϵ. Four different regions can be identified:

(I) $R_{\mathrm{RN}}(\bar{r})$ has 3 positive real zeros $r_1 < r_2 < r_3$ with $R_{\mathrm{RN}}(\bar{r}) \geq 0$ for $r_1 \leq \bar{r} \leq r_2$ and $r_3 \leq \bar{r}$. Possible orbit types: flyby and bound orbits.

(II) $R_{\mathrm{RN}}(\bar{r})$ has 1 positive real zero r_1 with $R_{\mathrm{RN}}(\bar{r}) \geq 0$ for $r_1 \leq \bar{r}$. Possible orbit types: flyby orbits.

(III) $R_{\mathrm{RN}}(\bar{r})$ has 2 positive real zeros r_1, r_2 with $R_{\mathrm{RN}}(\bar{r}) \geq 0$ for $r_1 \leq \bar{r} \leq r_2$. Possible orbit types: bound orbits.

(IV) $R_{\mathrm{RN}}(\bar{r})$ has 4 positive real zeros $r_1 < r_2 < r_3 < r_4$ with $R_{\mathrm{RN}}(\bar{r}) \geq 0$ for $r_1 \leq \bar{r} \leq r_2$ and $r_3 \leq \bar{r} \leq r_4$. Possible orbit types: two different bound orbits.

As in Schwarzschild space-time for light rays in black hole space-times with $\bar{Q}^2 \leq 1$ only regions (I) and (II) are possible. Furthermore, for a naked singularity with $1 < \bar{Q}^2 < \frac{9}{8}$ an additional part of region (II) appears below region (I). As $\bar{Q}^2$ grows, this part gets larger and, finally, at $\bar{Q}^2 = \frac{9}{8}$ region

3. Geodesics in spherically symmetric space-times

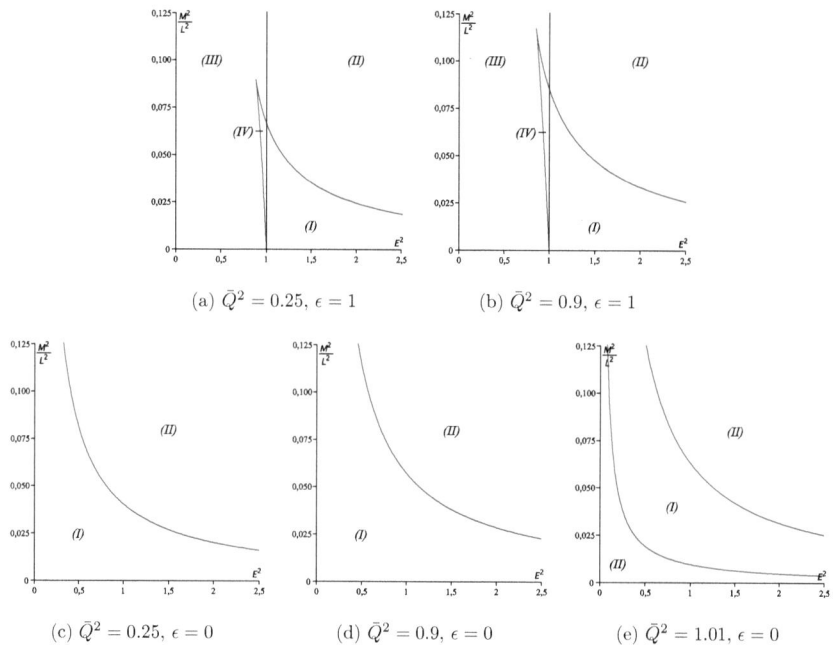

Figure 3.6: Regions of different types of geodesic motion in Reissner-Nordström space-time for different values of $\bar{Q}^2$ and ϵ. For null geodesics ($\epsilon = 0$) an example for a naked singularity with $1 < \bar{Q}^2 < \frac{9}{8}$ (see (e)) is included. If $\bar{Q}^2 \geq \frac{9}{8}$, region (I) vanishes and only region (II) is possible. For effective potentials see Fig. 3.7.

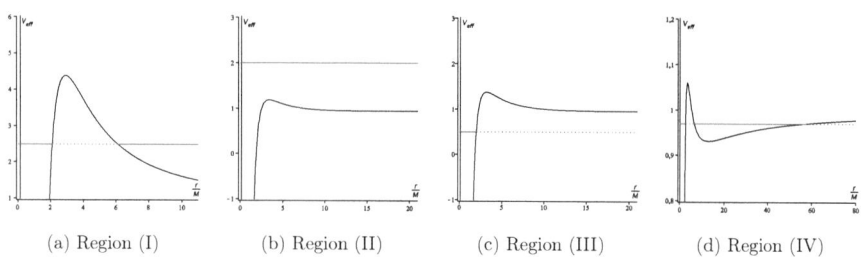

Figure 3.7: Effective potentials for different regions of geodesic motion in Reissner-Nordström space-time. The horizontal lines denote the squared energy parameter E^2.

3.2. Schwarzschild and Reissner-Nordström space-times

region	pos. zeros	range of $\bar{r}$	types of orbits
I	3	⊢―●――●――――→	flyby, bound
II	1	⊢―●――――――→	flyby
III	2	⊢―●――――●	bound
IV	4	⊢―●―●――●―●	2x bound

Table 3.2: Orbit types in Reissner-Nordström space-time. The second column gives the number of positive zeros of the polynomial R_{RN}. In the third column, the thick lines represent the range of orbits and turning points are shown by thick dots. The small vertical line denotes $r = 0$.

(I) vanishes completely and only region (II), i.e. flyby orbits, remains. For all regions of geodesic motion examples of effective potentials can be found in Fig. 3.7. A summary of possible orbit types can be found in Tab. 3.2.

Compared to the situation in Schwarzschild space-time, in each region there appears an additional real and positive zero r_1 (say) preventing test particles and light from falling into the singularity. This can be interpreted as a gravitational repulsion originating from the electric charge $\bar{Q}^2$ of the black hole. The additional zero r_1 lies always inside the Cauchy horizon at $\bar{r}_C = 1 - \sqrt{1 - \bar{Q}^2}$, what can be seen from the effective potential $V_{\text{eff}} = \frac{\Delta}{r^2}\left(\epsilon + \frac{L^2}{r^2}\right)$ defined in (3.1.5): because of $V_{\text{eff}} = 0$ at r_C and $V_{\text{eff}} = \infty$ at $r = 0$, V_{eff} takes all positive values between $r = 0$ and $r = r_C$ and, in particular, $V_{\text{eff}} = E^2$ for some $r \in (0, r_C)$. Thus, a flyby or bound orbit with the additional smallest positive real zero $r_1 < \bar{r}_C$ as periapsis crosses the Cauchy horizon, is reflected by the charge induced potential barrier and again crosses the Cauchy horizon in the opposite direction, thereby entering a new copy of the Reissner-Nordström spacetime. This can be inferred from the Carter-Penrose diagram of the Reissner-Nordström space-time shown, e.g., in [58] or [14]. Such a flyby orbit may be called a *two-world escape orbit*, see Fig. 3.8(b) for an example. For bound orbits, by proceeding further along its r-periodic motion, the particle or light ray again and again enters new copies of the Reissner-Nordström space-time within its analytic continuation. This may be called a *many-world bound orbit*, which can be seen, e.g., in Fig. 3.8(c).

Also, for increasing electric charge the regions (I) and (IV) gets larger. Thus, a pair $(E^2, \mathcal{L})$ which would be located in region (II) in the Schwarzschild case may be located in region (I) for nonvanishing $\bar{Q}^2$. Likewise, a pair $(E^2, \mathcal{L})$ which would be located in region (III) in a Schwarzschild space-time may be located in region (IV) in the Reissner-Nordström case. In region (IV) two different bound orbits are possible, one of which is a many-world orbit. Both of these orbits have the same periodicity as implied by the form of $r(\varphi)$ (see next subsection) given in terms of elliptic function, which may have one real period only.

3. Geodesics in spherically symmetric space-times

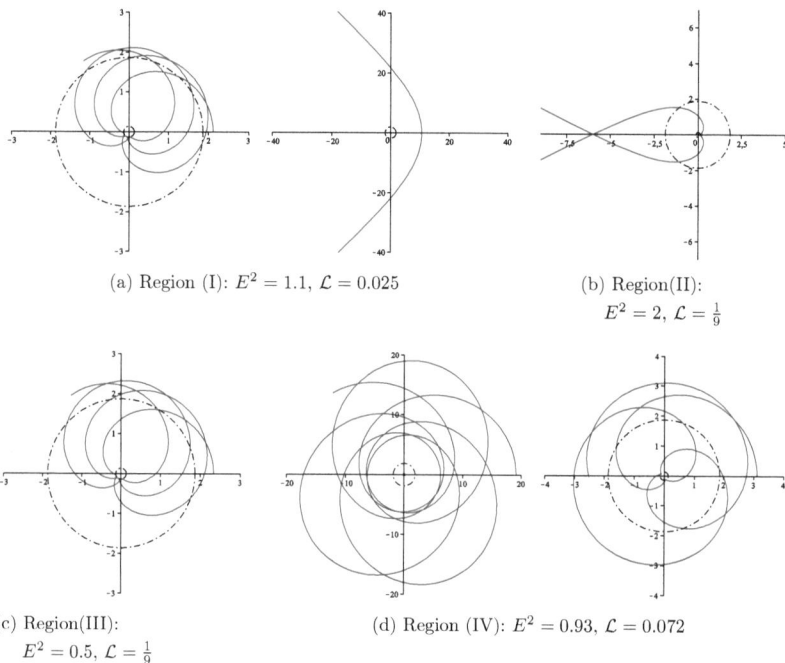

(a) Region (I): $E^2 = 1.1$, $\mathcal{L} = 0.025$

(b) Region(II): $E^2 = 2$, $\mathcal{L} = \frac{1}{9}$

(c) Region(III): $E^2 = 0.5$, $\mathcal{L} = \frac{1}{9}$

(d) Region (IV): $E^2 = 0.93$, $\mathcal{L} = 0.072$

Figure 3.8: Timelike geodesics in Reissner-Nordström space-time with $\bar{Q}^2 = \frac{1}{4}$ for every region of orbit types. The plots are in units of M and dashed lines mark the horizons. All bound orbits with the exception of the bound orbit on the left in (d) cross the inner Cauchy horizon just barely and, thus, are many-world bound orbits. Also, the flyby orbit in (b) is a two-world orbit.

In Figs. 3.8 and 3.9 sample geodesics are shown for each of the regions (I) to (IV), for both test particles and light.

Exceptional orbits appear for parameters $(E^2, \mathcal{L})$ which are located at the boundaries of regions (I) to (IV). These pairs of parameters correspond to $E^2 = 1$ or to multiple zeros of R_{RN}. In the latter case E^2 and $\mathcal{L}$ are given by Eq. (3.2.16) for test particles and by Eq.(3.2.17) for light. For $\epsilon = 0$ a multiple zero of $R_{\text{RN}}(\bar{r})$ can only exist if $\bar{Q}^2 \leq \frac{9}{8}$ and is then always located at $\bar{r} = \frac{3}{2} + \frac{1}{2}\sqrt{9 - 8\bar{Q}^2}$, which is a saddle point for a naked singularity and $\bar{Q}^2 = \frac{9}{8}$, and else a minimum corresponding to unstable circular orbits. In particular, for light in black hole space-times with $\bar{Q}^2 \leq 1$ stable circular

3.2. Schwarzschild and Reissner-Nordström space-times

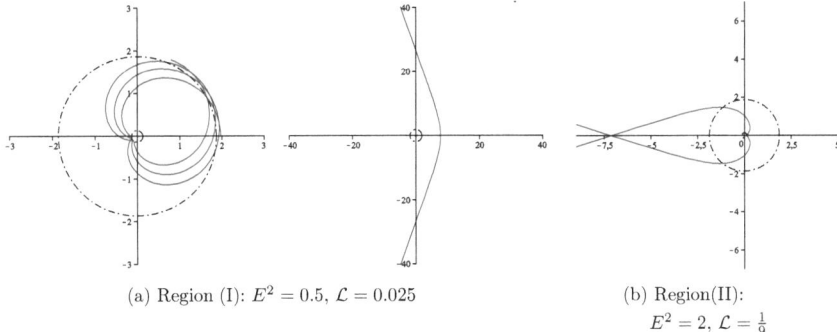

(a) Region (I): $E^2 = 0.5$, $\mathcal{L} = 0.025$

(b) Region(II): $E^2 = 2$, $\mathcal{L} = \frac{1}{9}$

Figure 3.9: Null geodesics in Reissner-Nordström space-time with $\bar{Q}^2 = \frac{1}{4}$ for every region of orbit types. The plots are in units of M and the dashed lines mark the horizon. The bound orbit in (a) is a many-world orbit and the flyby orbit in (b) a two-world orbit.

orbits do not exist. For test particles a substitution of Eq. (3.2.16) into $\frac{d^2}{d\bar{r}^2} R_{\mathrm{RN}}(\bar{r})$ yields

$$\frac{d^2}{dx^2} R_{\mathrm{RN}}(x) = -2 \frac{x^3 - 6x^2 + 9\bar{Q}^2 x - 4\bar{Q}^4}{x^2(x - \bar{Q}^2)}. \tag{3.2.18}$$

This expressions has one or three real zero by Descartes' rule of signs of which only one zero, $\bar{r}_{tr}$ say, corresponds to finite values of E^2 and $\mathcal{L}$ greater than or equal to zero. As in Schwarzschild space-time, the double zero x is a minimum if $x < \bar{r}_{tr}$ and, thus, corresponds to unstable circular orbits. If $x > \bar{r}_{tr}$ this is a maximum and corresponds to a stable circular orbit. At $\bar{r}_{tr}$ the polynomial R_{RN} has a triple zero, which is a saddle point. This is the innermost stable circular orbit. For $E^2 = \epsilon = 1$ the degree of R_{RN} decreases to 3 and R_{RN} has a double zero x corresponding to a circular orbit if

$$\mathcal{L}(x) = \frac{x-1}{x(3x + x(\sqrt{x} - 2))}, \quad \bar{Q}^2(x) = (2 \pm \sqrt{x})x, \tag{3.2.19}$$

which reduces to the Schwarzschild case for $x = 4$. If $E^2 = \epsilon = 1$ and R_{RN} has only simple zeros the possible orbit types can be inferred in the same way as for $E^2 \neq 1$.

3.2.2.2 Analytical solution of geodesic equations

As in Schwarzschild space-time the differential equation (3.2.14),

$$\left(\frac{d\bar{r}}{d\varphi}\right)^2 = R_{\mathrm{RN}}(\bar{r}) = (E^2 - \epsilon)\mathcal{L}\bar{r}^4 + 2\epsilon\mathcal{L}\bar{r}^3 - (\epsilon\mathcal{L}\bar{Q}^2 + 1)\bar{r}^2 + 2\bar{r} - \bar{Q}^2, \tag{3.2.14}$$

is of elliptic type if R_{RN} has only simple zeros but can be solved in terms of elementary functions if R_{RN} has multiple zeros. In the latter case the analytical solution of Eq. (3.2.14) can be found in [14, 31], for example. In the following, it is assumed that R_{RN} has only simple zeros.

3. Geodesics in spherically symmetric space-times

With the new variable $u = M/r$ Eq. (3.1.3) reads for Reissner-Nordström space-times

$$\left(\frac{du}{d\varphi}\right)^2 = -\bar{Q}^2 u^4 + 2u^3 - (1 + \epsilon\bar{Q}^2 \mathcal{L})u^2 + 2\epsilon\mathcal{L}u + \mathcal{L}(E^2 - \epsilon) \tag{3.2.20}$$

with the dimensionless parameters introduced in Eq. (3.2.13)

$$\mathcal{L} := \frac{M^2}{L^2}, \quad \bar{Q} = \frac{Q}{M}.$$

However, for the conversion of Eq. (3.2.14) to the Weierstrass form (2.1.5) the standard substitution $r = \frac{1}{\xi} + r_{\rm RN}$ for a zero $r_{\rm RN}$ of $R_{\rm RN}(r)$ is more convenient because it transforms the right hand side of (3.2.14) to a polynomial of degree 3,

$$\left(\frac{d\xi}{d\varphi}\right)^2 = \sum_{j=0}^{3} a_j \xi^j, \quad a_j = \frac{1}{(4-j)!} \frac{d^{(4-j)} R_{\rm RN}}{d\bar{r}^{4-j}} (r_{\rm RN}). \tag{3.2.21}$$

Finally, an additional substitution $\xi = \frac{1}{a_3}\left(4y - \frac{a_2}{3}\right)$ casts (3.2.21) in the Weierstrass form

$$\left(\frac{dy}{d\varphi}\right)^2 = 4y^3 - g_2 y - g_3, \tag{3.2.22}$$

where g_2, g_3 are given by (2.1.11)

$$g_2 = \frac{1}{16}\left(\frac{4}{3}a_2^2 - 4a_1 a_3\right),$$

$$g_3 = \frac{1}{16}\left(\frac{1}{3}a_1 a_2 a_3 - \frac{2}{27}a_2^3 - a_0 a_3^2\right).$$

The analytical solution of Eq. (3.1.3) for Reissner-Nordström space-time is then given by

$$r(\varphi) = \frac{1}{\xi(\varphi)} + r_{\rm RN} = \frac{a_3}{4\wp(\varphi - \varphi_{in}) - \frac{a_2}{3}} + r_{\rm RN}, \tag{3.2.23}$$

where

$$\varphi_{in} = \varphi_0 + \int_{y_0}^{\infty} \frac{dz}{\sqrt{4z^3 - g_2 z - g_3}}, \quad y_0 = \frac{1}{4}\left(\frac{a_3}{r_0 - r_{\rm RN}} + \frac{a_2}{3}\right), \tag{3.2.24}$$

depends only on the initial values φ_0 and r_0. In Figs. 3.8 and 3.9 this solution was used to create the examples of timelike and null geodesics for each region of different types of orbits identified in the foregoing subsection.

3.3 Schwarzschild- and Reissner-Nordström-(anti-)de Sitter space-times

In this section the results of Sec. 3.2 are generalized to the case of a nonvanishing cosmological constant, i.e. the geodesic equation (3.0.3)

$$0 = \frac{d^2 x^\mu}{ds^2} + \left\{{}^{\mu}_{\rho\sigma}\right\} \frac{dx^\rho}{ds} \frac{dx^\sigma}{ds}$$

in Schwarzschild-de Sitter and Reissner-Nordström-de Sitter space-times given by (3.0.7) and (3.0.8) is considered. These space-times are characterized by the mass M, the cosmological constant Λ, and in the case of Reissner-Nordström-de Sitter by the electric charge Q of the black hole. Background informations on de Sitter space-times and the cosmological constant can be found in [59] and [60], respectively. Analogously the the case of a vanishing cosmological constant the geodesic equation can be reduced to the ordinary differential equations (3.1.2)-(3.1.4)

$$\left(\frac{dr}{d\varphi}\right)^2 = R(r) = \frac{r^4}{L^2}\left(E^2 - \frac{\Delta}{r^2}\left(\epsilon + \frac{L^2}{r^2}\right)\right), \tag{3.1.3}$$

$$\left(\frac{dr}{ds}\right)^2 = E^2 - \frac{\Delta}{r^2}\left(\epsilon + \frac{L^2}{r^2}\right), \tag{3.1.2}$$

$$\left(\frac{dr}{dt}\right)^2 = \frac{1}{E^2}\frac{\Delta^2}{r^4}\left(E^2 - \frac{\Delta}{r^2}\left(\epsilon + \frac{L^2}{r^2}\right)\right). \tag{3.1.4}$$

In the following these equation are treated for the two classes of space-times separately along the lines of [61, 62, 63, 64].

3.3.1 Geodesics in Schwarzschild-(anti-)de Sitter space-times

3.3.1.1 Types of orbits

Analogously to the case of a vanishing cosmological constant the types of orbits in Schwarzschild-de Sitter space-time are encoded in the right hand side of Eq. (3.1.3)

$$\left(\frac{dr}{d\varphi}\right)^2 = R(r) = \frac{r^4}{L^2}\left(E^2 - \frac{\Delta_{\text{SdS}}}{r^2}\left(\epsilon + \frac{L^2}{r^2}\right)\right), \tag{3.1.3}$$

where $\Delta_{\text{SdS}} = r^2 - 2Mr - \frac{1}{3}\Lambda r^4$. With the dimensionless quantities

$$\bar{r} := \frac{r}{M}, \quad \mathcal{L} := \frac{M^2}{L^2}, \quad \bar{\Lambda} = \frac{1}{3}\Lambda M^2, \tag{3.3.1}$$

Eq. (3.1.3) can be written as

$$\left(\frac{d\bar{r}}{d\varphi}\right)^2 = \epsilon\bar{\Lambda}\mathcal{L}\bar{r}^6 + ((E^2 - \epsilon)\mathcal{L} + \bar{\Lambda})\bar{r}^4 + 2\epsilon\mathcal{L}\bar{r}^3 - \bar{r}^2 + 2\bar{r} =: R_{\text{SdS}}(\bar{r}). \tag{3.3.2}$$

A special feature of space-times with a nonvanishing and small positive cosmological constant is, compared to Schwarzschild space-times, the additional cosmological horizon located at the largest of the two positive real zeros of $\Delta_{\text{SdS}} = M^2(-\bar{\Lambda}\bar{r}^4 + \bar{r}^2 - 2\bar{r})$. More precisely, this cosmological horizon appears for $\bar{\Lambda} < \frac{1}{27}$ (or $\Lambda < \frac{1}{9M^2}$) and induces a potential barrier at which the geodesic motion may be reflected. For $\bar{\Lambda} = \frac{1}{27}$ there is only one horizon at $r = 3M$, but for larger values of $\bar{\Lambda}$ the singularity is naked. If the cosmological constant is negative, there is one horizon as in Schwarzschild space-time.

3. Geodesics in spherically symmetric space-times

Eq. (3.3.2) implies that $R_{\text{SdS}} \geq 0$ is a necessary condition for the existence of a geodesic and, thus, that the real and positive zeros of R_{SdS} are extremal values of the geodesic motion. As $\bar{r} = 0$ is a zero of R_{SdS} for all values of the parameters, it is neglected in the following analysis and

$$R^*_{\text{SdS}}(\bar{r}) = \epsilon \bar{\Lambda} \mathcal{L} \bar{r}^5 + ((E^2 - \epsilon)\mathcal{L} + \bar{\Lambda}) \bar{r}^3 + 2\epsilon \mathcal{L} \bar{r}^2 - \bar{r} + 2 \qquad (3.3.3)$$

is considered instead.

It can be shown that R^*_{SdS} has no more than four real and positive zeros by decomposing the polynomial R^*_{SdS} into its (in general complex) zeros $R^*_{\text{SdS}}(\bar{r}) = (\bar{r} - \bar{r}_1)(\bar{r} - \bar{r}_2)(\bar{r} - \bar{r}_3)(\bar{r} - \bar{r}_4)(\bar{r} - \bar{r}_5)$. Multiplication and comparison of the coefficients of $\bar{r}^4$ yields

$$-\sum_{i=1}^{5} \bar{r}_i = 0, \qquad (3.3.4)$$

what contradicts the assumption that all zeros are real and positive. Therefore, in any case there are at most four real positive zeros.

The behavior of R^*_{SdS} at $\bar{r} = \infty$ depends on the sign of $\bar{\Lambda}$ if $\epsilon = 1$ and on the sign of $E^2 \mathcal{L} + \bar{\Lambda}$ if $\epsilon = 0$. Let $e_1 < \ldots < e_n$ denote the positive and real zeros of R^*_{SdS}. If $\lim_{\bar{r} \to \infty} R^*_{\text{SdS}}(\bar{r}) = \infty$ then it follows that the physically acceptable regions are given by $[0, e_1], [e_2, e_3], \ldots, [e_n, \infty]$ if n is even and by $[e_1, e_2], \ldots, [e_n, \infty]$ if n is odd. However, if $\lim_{\bar{r} \to \infty} R^*_{\text{SdS}}(\bar{r}) = -\infty$ the physically acceptable regions are given by $[e_1, e_2], \ldots, [e_{n-1}, e_n]$ if n is even and by $[0, e_1], [e_2, e_3], \ldots, [e_{n-1}, e_n]$ if n is odd.

If for a given set of parameters $\bar{\Lambda}$, ϵ, E^2, and $\mathcal{L}$ the polynomial R^*_{SdS} has n positive and real zeros then for varying E^2 and $\mathcal{L}$ this number can only change if two zeros merge to one. (A positive zero can not become negative as $R^*_{\text{SdS}}(0) = 2$ and R^*_{SdS} is continuous in E^2 and $\mathcal{L}$.) A merger of two zeros happens at $\bar{r} = x$ if and only if

$$R^*_{\text{SdS}}(\bar{r}) = (\bar{r} - x)^2 (a_3 \bar{r}^3 + a_2 \bar{r}^2 + a_1 \bar{r} + a_0) \qquad (3.3.5)$$

for some real constants a_i. By a comparison of coefficients we can solve the resulting 6 equations for E^2 and $\mathcal{L}$ dependent on ϵ and $\bar{\Lambda}$. For $\epsilon = 1$ this yields

$$E^2(x) = \frac{(\bar{\Lambda} x^3 - (x - 2))^2}{x(x - 3)}, \quad \mathcal{L}(x) = \frac{x - 3}{x^2 (1 - \bar{\Lambda} x^3)}, \qquad (3.3.6)$$

where x is the position of the double zero, and for $\epsilon = 0$

$$\mathcal{L} = \frac{1}{27 E^2} - \frac{\bar{\Lambda}}{E^2}. \qquad (3.3.7)$$

In Fig. 3.10 the results of this analysis are shown for both test particles ($\epsilon = 1$) and light ($\epsilon = 0$) for a small positive cosmological constant $\Lambda > 0$. Three different regions of geodesic motion can be identified

3.3. S- and RN-(anti-)dS space-times

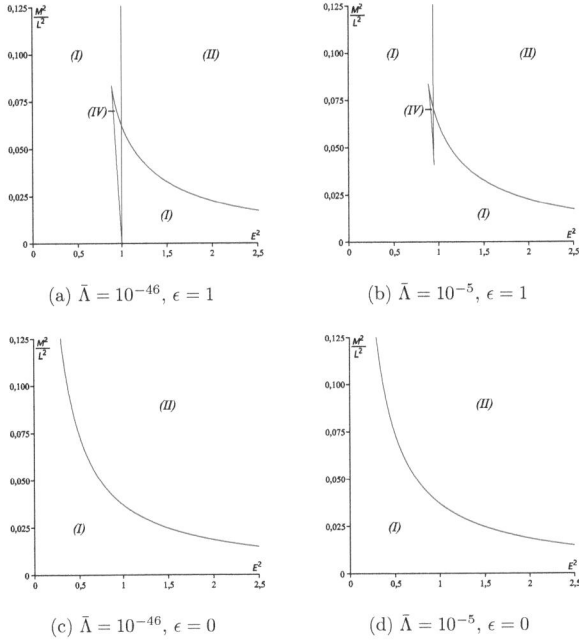

Figure 3.10: Regions of different types of geodesic motion in Schwarzschild-de Sitter space-time for varying $\bar{\Lambda}$ and ϵ. For effective potentials see Fig. 3.11.

(I) $R^*_{\text{SdS}}(\bar{r})$ has 2 positive real zeros $r_1 < r_2$ with $R_{\text{SdS}}(\bar{r}) \geq 0$ for $0 \leq \bar{r} \leq r_1$ and $r_2 \leq \bar{r}$. Possible orbit types: flyby and terminating bound orbits.

(II) $R^*_{\text{SdS}}(\bar{r})$ has 0 positive real zeros and $R_{\text{SdS}}(\bar{r}) \geq 0$ for positive $\bar{r}$. Possible orbit types: terminating escape orbits.

(IV) $R^*_{\text{SdS}}(\bar{r})$ has 4 positive real zeros $r_i < r_{i+1}$ with $R_{\text{SdS}}(\bar{r}) \geq 0$ for $0 \leq \bar{r} \leq r_1$, $r_2 \leq \bar{r} \leq r_3$, and $r_4 \leq \bar{r}$. Possible orbit types: flyby, bound, and terminating bound orbits.

For light rays only regions (I) and (II) appear. Sample effective potentials for each of the regions (I),(II), and (IV) are shown in Fig. 3.11.

Now let us compare the geodesic motion of test particles ($\epsilon = 1$) with the case $\bar{\Lambda} = 0$. From Figs. 3.2 and 3.10 it is obvious that at the left of $E^2 = 1$ the regions significantly changed. Region (I) absorbed the whole parameter space which was located in region (III) for $\bar{\Lambda} = 0$. This means that left of $E^2 < 1$ there is now (at least) one more positive real zero for each pair $(E^2, \mathcal{L})$. In particular,

3. Geodesics in spherically symmetric space-times

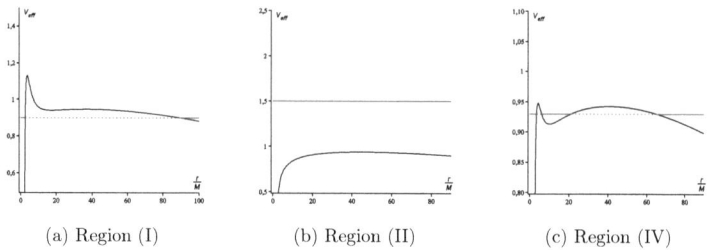

(a) Region (I) (b) Region (II) (c) Region (IV)

Figure 3.11: Effective potentials for the different regions of geodesic motion in Schwarzschild-de Sitter space-time. The horizontal lines denote the squared energy parameter E^2.

region	pos. zeros	range of $\bar{r}$	types of orbits
I	2	⊢―●――●―→	flyby, terminating bound
II	0	⊢―――――→	terminating escape
IV	4	⊢―●――●――●―→	flyby, bound, terminating bound

Table 3.3: Orbit types in Schwarzschild-de Sitter space-time for a small cosmological constant. The second column gives the number of positive zeros of the polynomial $R^*_{\text{SdS}}(\bar{r})$. In the third column, the thick lines represent the range of orbits and turning points are shown by thick dots. The small vertical line denotes $\bar{r} = 0$.

this implies that a particle which for $\bar{\Lambda} = 0$ was located in region (III) may now reach infinity. For $E^2 < 1$ there are also flyby orbits which, coming from infinity, are reflected at the potential barrier induced by the small positive cosmological constant, see Fig. 3.12. A test particle with $E^2 = 1 - \delta$ for a very small $\delta > 0$, which belonged to region (IV) or (III) for $\bar{\Lambda} = 0$, now switched to region (I) or (II) depending on its $\mathcal{L}$ value, see Fig. 3.13. In general, all regions for $\bar{\Lambda} > 0$ are a little bit shifted compared to $\bar{\Lambda} = 0$. Thus, every pair $(E^2, \mathcal{L})$ which was located near a boundary for $\bar{\Lambda} = 0$ may switch to another region. For every region, examples of timelike geodesics can be found in Fig. 3.12. Each of these plots was created using the analytical solution of $r(\varphi)$ derived in the next subsection. A summary of possible orbit types can be found in Tab. 3.3.

For a large positive cosmological constant the region (IV) will disappear, that is, no bound orbits exist. This is clear from the following: The boundary of region (IV) is defined by two corners, which correspond to the triple zeros of R^*_{SdS}. This means that the region will vanish if triple zeros are not possible. With an analysis similar to (3.3.5) it can be shown that triple zeros appear at

$$\mathcal{L} = \frac{4x - 15}{3x^2}, \quad E^2 = \frac{16(x^3 - 9x^2 + 27x - 27)}{x(4x - 15)^2}, \quad \bar{\Lambda} = \frac{x - 6}{x^3(4x - 15)}, \quad (3.3.8)$$

3.3. S- and RN-(anti-)dS space-times

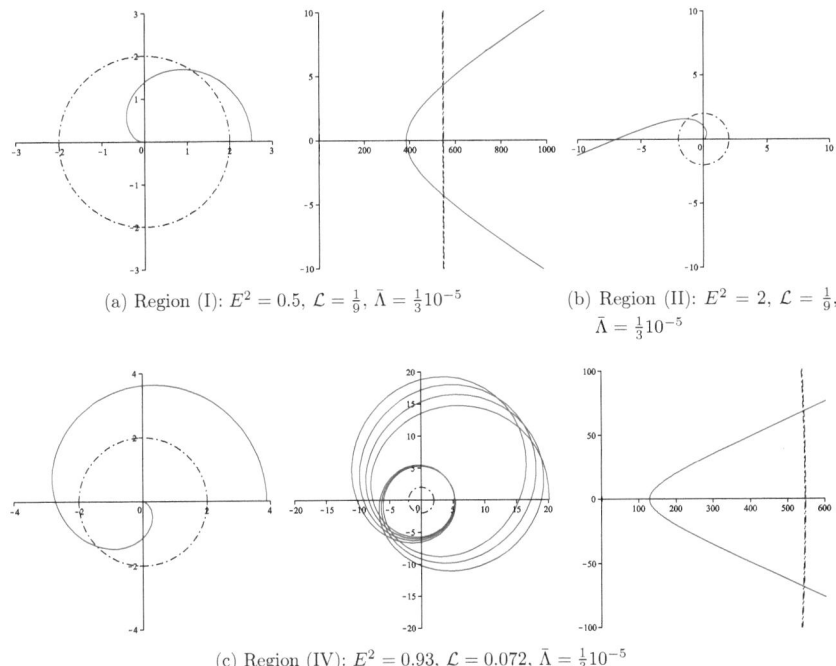

(a) Region (I): $E^2 = 0.5$, $\mathcal{L} = \frac{1}{9}$, $\bar{\Lambda} = \frac{1}{3}10^{-5}$

(b) Region (II): $E^2 = 2$, $\mathcal{L} = \frac{1}{9}$, $\bar{\Lambda} = \frac{1}{3}10^{-5}$

(c) Region (IV): $E^2 = 0.93$, $\mathcal{L} = 0.072$, $\bar{\Lambda} = \frac{1}{3}10^{-5}$

Figure 3.12: Timelike geodesics in Schwarzschild-de Sitter space-time for the different regions of orbit types. The plots are in units of M and dashed lines mark horizons. The flyby orbits in (a) and (c) are reflected at the potential barrier induced by the positive cosmological constant

where x is the position of the triple zero. It can be shown that $\mathcal{L} \geq 0$, $E^2 \geq 0$, and $\bar{\Lambda} \geq 0$ in (3.3.8) are simultaneously fulfilled if and only if $x \geq 6$. The maximum value for $\bar{\Lambda}$ from (3.3.8) and $x \geq 6$ is given by $x = \frac{15}{2}$ what corresponds to $\bar{\Lambda} = \frac{4}{16875} \approx 0.00024$. This means that for larger $\bar{\Lambda}$ triple zeros are not possible (or correspond to negative $\mathcal{L}$ or E^2) and, thus, that region (IV) vanishes for such large $\bar{\Lambda}$.

The types of orbits for null geodesics, i.e. for light rays, are not essentially influenced by a nonvanishing cosmological constant. It will be shown in the next subsection that null geodesics depend on one parameter only, which is given by $p_{\text{light}} := \mathcal{L}E^2 + \bar{\Lambda}$, in both Schwarzschild and Schwarzschild-de Sitter space-time. This means that two light rays, one moving in Schwarzschild and one in Schwarzschild-de Sitter space-time, with identical p_{light} can not be distinguished. This becomes even more obvious if p_{light} is expressed in terms of the point $\bar{r}_p$ of closest approach to the

3. Geodesics in spherically symmetric space-times

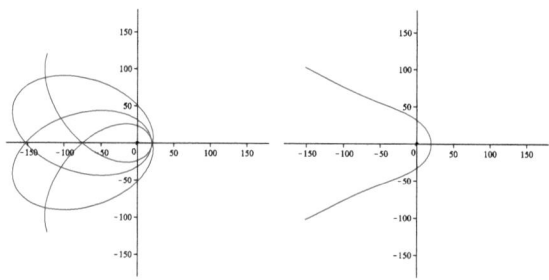

(a) $\epsilon = 1$, $E^2 = 0.99$, $\mathcal{L} = 0.025$. Left: $\bar{\Lambda} = 0$, region(IV)
right: $\bar{\Lambda} = \frac{1}{3} \times 10^{-5}$, region(I)

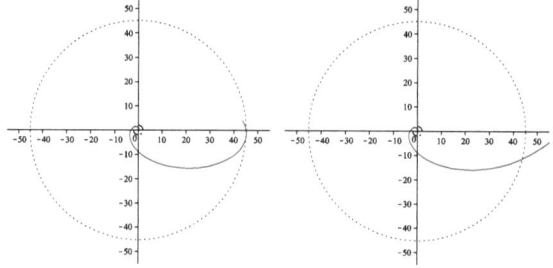

(b) $\epsilon = 1$, $E^2 = 0.96$, $\mathcal{L} = \frac{1}{9}$. Left: $\bar{\Lambda} = 0$, region (III)
right: $\bar{\Lambda} = \frac{1}{3} \times 10^{-5}$, region (II)

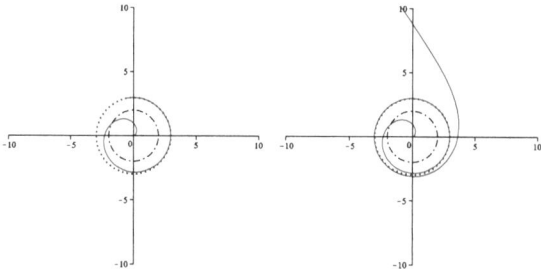

(c) $\epsilon = 0$, $\mathcal{L}E^2 = \frac{1}{27} - \frac{1}{6}10^{-5}$. Left: $\bar{\Lambda} = 0$, region (I)
right: $\bar{\Lambda} = \frac{1}{3} \times 10^{-5}$, region (II)

Figure 3.13: Comparison between orbits in Schwarzschild and Schwarzschild-de Sitter space-time. In (b) and (c) the maximal r in Schwarzschild space-time are $r_{\max} \approx 45.25$ and $r_{\max} \approx 2.99$ indicated by the dotted lines. Dashed lines mark horizons.

gravitational center. This is given by $\frac{d\bar{r}}{d\varphi} = 0$ and, therefore,

$$p_{\text{light}} \bar{r}_p^4 - \bar{r}_p^2 + 2\bar{r}_p = 0 \quad \Leftrightarrow \quad p_{\text{light}} = \frac{1}{\bar{r}_p^2} - \frac{2}{\bar{r}_p^3}. \tag{3.3.9}$$

In terms of this parameter the geodesic equation (3.3.2) for $\epsilon = 0$ reads

$$\left(\frac{d\bar{r}}{d\varphi}\right)^2 = \left(\frac{1}{\bar{r}_p^2} - \frac{2}{\bar{r}_p^3}\right)\bar{r}^4 - \bar{r}^2 + 2\bar{r}. \tag{3.3.10}$$

However, if $\mathcal{L}E^2$ is fixed and a cosmological constant is 'switched on' a null geodesic is heavily changed if $\frac{1}{27} - \bar{\Lambda} \leq \mathcal{L}E^2 \leq \frac{1}{27}$ as can be read of (3.3.7).

In a recent paper [65] Rindler and Ishak discussed the light deflection in a Schwarzschild-de Sitter space-time. Though the equation of motion is the same as in Schwarzschild space-time for identical periapsis $\bar{r}_p$, they showed that the measuring process for angles reintroduces the effect of the cosmological constant. According to their scheme, the exact angle between the radial direction and the spatial direction of the light ray is now given by

$$\tan \psi = \frac{\sqrt{1 - \frac{2M}{r(\varphi)} - \frac{1}{3}\Lambda r^2(\varphi)}}{\left|\sqrt{\frac{r^2(\varphi)}{r_p^2}\left(1 - \frac{2M}{r_p}\right) - \left(1 - \frac{2M}{r(\varphi)}\right)}\right|}, \tag{3.3.11}$$

where $r_p = M\bar{r}_p$ and $r(\varphi)$ is the solution of (3.3.2) for $\epsilon = 0$ derived in the next section, see (3.3.15). This now is valid for all light rays, not only for those rays showing a small deflection as discussed in [65].

Let us also discuss exceptional orbits for a small positive Λ, which appear at parameter values of E^2 and $\mathcal{L}$ which are located on the boundaries of the regions (I), (II), and (IV). These parameter values are given by Eq. (3.3.6) for particles and Eq. (3.3.7) for light. For particles a substitution of (3.3.6) into $\frac{d^2}{d\bar{r}^2} R^*_{\text{SdS}}(\bar{r})$ yields

$$\frac{d^2}{dx^2} R^*_{\text{SdS}}(x) = -\frac{2(x - 6 - \bar{\Lambda}x^3(4x - 15))}{x^2(1 - \bar{\Lambda}x^3)}. \tag{3.3.12}$$

This expression has a maximum of two positive zeros (which are triple zeros of R^*_{SdS}) by Descartes' rule of signs, say $\bar{r}_1 \leq \bar{r}_2$. Together with Eq. (3.3.8) and the following discussion it can be inferred that $6 \leq \bar{r}_1 \leq \frac{15}{2} \leq \bar{r}_2$, where $\bar{r}_1 = 6$ is the Schwarzschild case $\bar{\Lambda} = 0$, cf. Eq. (3.2.7). For small $\bar{\Lambda}$ the triple zero $\bar{r}_1$ corresponds to the corner in the boundary of region (IV) at $E^2(\bar{r}_1) \approx \frac{8}{9}$ and $\mathcal{L}(\bar{r}_1) \approx \frac{1}{12}$. This is the innermost stable circular orbit in Schwarzschild-de Sitter space-time. The larger triple zero $\bar{r}_2$ corresponds to the other corner in the boundary of region (IV), see Fig. 3.10. A double zero $x < \bar{r}_1$ or $\bar{r}_2 < x$ is a minimum and corresponds to an unstable circular orbit whereas a double zero $\bar{r}_1 < x < \bar{r}_2$ is a maximum and corresponds to a stable circular orbit. Thus, $\bar{r}_2$ can be called the outermost stable circular orbit. For light rays double zeros can only be located at $x = 3$ (or $r = 3M$) as in Schwarzschild space-time. This is always a minimum and, thus, corresponds to an unstable circular orbit.

3.3.1.2 Analytical solution of geodesic equations

The structure of Eq. (3.1.3) in Schwarzschild-(anti-)de Sitter space-time,

$$\left(\frac{dr}{d\varphi}\right)^2 = R(r) = \frac{r^4}{L^2}\left(E^2 - \frac{\Delta_{\text{SdS}}}{r^2}\left(\epsilon + \frac{L^2}{r^2}\right)\right), \qquad (3.1.3)$$

where $\Delta_{\text{SdS}} = r^2 - 2Mr - \frac{1}{3}\Lambda r^4$ is very different for null ($\epsilon = 0$) and timelike ($\epsilon = 1$) geodesics. In the case of $\epsilon = 0$ Eq. (3.1.3) can be solved analogously to Schwarzschild space-time, but for timelike geodesics the solution of Eq. (3.1.3) can be found as a limiting case of the solution of Jacobi's inversion problem in the case of genus $g = 2$. Thus, this problem and the corresponding mathematical foundation are first specialized to the case of $g = 2$ before the limiting process is performed.

As in Schwarzschild space-time a new variable $u = M/r$ is introduced, which yields

$$\left(\frac{du}{d\varphi}\right)^2 = 2u^3 - u^2 + 2\epsilon\mathcal{L}u + \left(\mathcal{L}(E^2 - \epsilon) + \bar{\Lambda}\right) + \epsilon\mathcal{L}\bar{\Lambda}\frac{1}{u^2} \qquad (3.3.13)$$

with the dimensionless parameters introduced in (3.3.1)

$$\mathcal{L} = \frac{M^2}{L^2}, \qquad \bar{\Lambda} = \frac{1}{3}\Lambda M^2.$$

Null geodesics For $\epsilon = 0$ Eq. 3.3.13 is of elliptic type and can be solved in the same way as the geodesic equation in Schwarzschild space-time. With the standard substitution $u = 2y + \frac{1}{6}$ Eq. (3.3.13) can be transformed to the Weierstrass form (2.1.5) with

$$\begin{aligned} g_2 &= \frac{1}{12}, \\ g_3 &= \frac{1}{216} - \frac{1}{4}\left(\mathcal{L}E^2 + \bar{\Lambda}\right). \end{aligned} \qquad (3.3.14)$$

Here it becomes obvious that the solution of (3.3.13) depends only on one parameter $p_{\text{light}} = \mathcal{L}E^2 + \bar{\Lambda}$ as in Schwarzschild space-time, where this parameter is defined by $p_{\text{light}} = \mathcal{L}E^2$. The analytical solution of (3.3.13) for $\epsilon = 0$ is then given by

$$r(\varphi) = \frac{M}{2\wp(\varphi - \varphi_{in}) + \frac{1}{6}}, \qquad (3.3.15)$$

where

$$\varphi_{in} = \varphi_0 + \int_{y_0}^{\infty} \frac{dz}{\sqrt{4y^3 - g_2 - g_3}}, \qquad y_0 = \frac{1}{2}\left(\frac{M}{r_0} - \frac{1}{6}\right) \qquad (3.3.16)$$

depends only on the initial values φ_0 and r_0.

3.3. S- and RN-(anti-)dS space-times

Timelike geodesics For $\epsilon = 1$ Eq. (3.3.13) should be rewritten as

$$\left(u\frac{du}{d\varphi}\right)^2 = 2u^5 - u^4 + 2\mathcal{L}u^3 + \left(\mathcal{L}(E^2-1) + \bar\Lambda\right)u^2 + \mathcal{L}\bar\Lambda =: P_{\text{SdS}}(u). \tag{3.3.17}$$

A separation of variables in (3.3.17) yields

$$\varphi - \varphi_0 = \int_{u_0}^{u} \frac{u\,du}{\sqrt{P_{\text{SdS}}(u)}}, \tag{3.3.18}$$

where $u_0 = u(\varphi_0)$. In solving integral (3.3.18) there are two major issues which have to be addressed. First, the integrand is not well defined in the complex plane because of the two branches of the square root. Second, the solution $u(\varphi)$ should not depend on the integration path. If γ denotes some closed integration path and

$$\oint_\gamma \frac{u\,du}{\sqrt{P_{\text{SdS}}(u)}} = \omega \tag{3.3.19}$$

this means that

$$\varphi - \varphi_0 - \omega = \int_{u_0}^{u} \frac{u\,du}{\sqrt{P_{\text{SdS}}(u)}} \tag{3.3.20}$$

should be valid, too. Hence, the solution $u(\varphi)$ of our problem has to fulfill

$$u(\varphi) = u(\varphi - \omega) \tag{3.3.21}$$

for every $\omega \neq 0$ obtained from an integration (3.3.19). These two issues can be solved if we consider Eq. (3.3.18) to be defined on the Riemann surface $X := \{(x,y) \in \mathbb{C}^2 \,|\, y^2 = P_{\text{SdS}}(x)\}$ of genus $g = 2$. There are 4 independent closed paths on X and 2 independent holomorphic differentials (cf. Eqs. (2.2.2) and (2.2.3))

$$dz_1 := \frac{dx}{\sqrt{P_{\text{SdS}}(x)}}, \qquad dz_2 := \frac{x\,dx}{\sqrt{P_{\text{SdS}}(x)}}, \tag{3.3.22}$$

where $\mathcal{L} = \frac{M^2}{L^2}$ is defined in (3.3.1).

Jacobi's inversion problem We want to identify now Eq. (3.3.18) as a part of Jacobi's inversion problem (2.4.4). To start with, we rewrite the inversion problem in the form

$$\begin{aligned}\phi_1 &= \int_\infty^{u_1} \frac{dx}{\sqrt{P_{\text{SdS}}(x)}} + \int_\infty^{u_2} \frac{dx}{\sqrt{P_{\text{SdS}}(x)}}, \\ \phi_2 &= \int_\infty^{u_1} \frac{x\,dx}{\sqrt{P_{\text{SdS}}(x)}} + \int_\infty^{u_2} \frac{x\,dx}{\sqrt{P_{\text{SdS}}(x)}},\end{aligned} \tag{3.3.23}$$

where

$$\vec\phi = \vec\varphi - 2\int_{u_0}^{\infty} d\vec{z}. \tag{3.3.24}$$

3. Geodesics in spherically symmetric space-times

Note that the right-hand side of (3.3.23) is exactly $\vec{A}_\infty(\vec{u})$, the image of the Abel map defined in Eq. (2.4.2), i.e. (3.3.23) can in short be written as $\vec{\phi} = \vec{A}_\infty(\vec{u})$. We use the obvious identity (compare [51])

$$u_1 = \lim_{u_2 \to \infty} \frac{u_1 u_2}{u_1 + u_2} \tag{3.3.25}$$

and insert the solution of Jacobi's inversion problem (2.4.8). Then

$$\begin{aligned} u_1 &= -\lim_{u_2 \to \infty} \frac{\wp_{12}(\vec{\phi})}{\wp_{22}(\vec{\phi})} \\ &= \lim_{u_2 \to \infty} \frac{\sigma(\vec{\phi})\sigma_{12}(\vec{\phi}) - \sigma_1(\vec{\phi})\sigma_2(\vec{\phi})}{\sigma_2^2(\vec{\phi}) - \sigma\sigma_{22}(\vec{\phi})} \\ &= \frac{\sigma(\vec{\phi}_\infty)\sigma_{12}(\vec{\phi}_\infty) - \sigma_1(\vec{\phi}_\infty)\sigma_2(\vec{\phi}_\infty)}{\sigma_2^2(\vec{\phi}_\infty) - \sigma(\vec{\phi}_\infty)\sigma_{22}(\vec{\phi}_\infty)}, \end{aligned} \tag{3.3.26}$$

where

$$\vec{\phi}_\infty = \lim_{u_2 \to \infty} \vec{\phi} = \lim_{u_2 \to \infty} \vec{A}_\infty(\vec{u}) = \vec{A}_\infty(\vec{u}_\infty) \tag{3.3.27}$$

with $\vec{u}_\infty = \begin{pmatrix} u_1 \\ \infty \end{pmatrix}$. Note that the definition of the Kleinian sigma function (2.4.5) and, hence, of the generalized Weierstrass functions (2.4.6) includes the vector of Riemann constant $\vec{K}_{x_0}$ with $x_0 = \infty$ in our case, which is given by

$$\vec{K}_\infty = \tau \begin{pmatrix} \frac{1}{2} \\ \frac{1}{2} \end{pmatrix} + \begin{pmatrix} 0 \\ \frac{1}{2} \end{pmatrix} \tag{3.3.28}$$

(see Eq. (2.3.8) or [49]).

Restriction to the theta divisor The limiting process $u_2 \to \infty$ in (3.3.25) also transfers Jacobi's inversion problem to the theta divisor $\Theta_{\vec{K}_{x_0}}$, which is defined as the set of zeros of $\vartheta(\cdot + \vec{K}_{x_0})$ (cf. (2.3.1)) and what simplifies Eq. (3.3.26) considerably. From

$$(2\omega)^{-1}\vec{\phi}_\infty = (2\omega)^{-1}\vec{A}_\infty(\vec{u}_\infty) = \int_\infty^{u_1} d\vec{v} \tag{3.3.29}$$

and the theorem

$$\vartheta \left[\begin{pmatrix} 1/2 \\ 1/2 \end{pmatrix}, \begin{pmatrix} 0 \\ 1/2 \end{pmatrix} \right](\vec{z}; \tau) = 0 \quad \Leftrightarrow \quad \exists x : \vec{z} = \int_\infty^x d\vec{v} \tag{3.3.30}$$

proven by Mumford [49], where $d\vec{v} = (2\omega)^{-1}d\vec{z}$ is the vector of normalized holomorphic differentials (cf. (2.2.6)), it follows that

$$0 = \vartheta \left[\begin{pmatrix} 1/2 \\ 1/2 \end{pmatrix}, \begin{pmatrix} 0 \\ 1/2 \end{pmatrix} \right]((2\omega)^{-1}\vec{\phi}_\infty; \tau). \tag{3.3.31}$$

Via Eq. (2.3.4) this is equivalent to

$$0 = \vartheta \left((2\omega)^{-1}\vec{\phi}_\infty + \tau \begin{pmatrix} 1/2 \\ 1/2 \end{pmatrix} + \begin{pmatrix} 0 \\ 1/2 \end{pmatrix}; \tau \right), \tag{3.3.32}$$

3.3. S- and RN-(anti-)dS space-times

which in turn implies

$$\sigma(\vec{\phi}_\infty) = 0 \tag{3.3.33}$$

using the definition (2.4.5) of the Kleinian sigma function. With this result we obtain from (3.3.26) the equation

$$u_1 = -\frac{\sigma_1(\vec{\phi}_\infty)}{\sigma_2(\vec{\phi}_\infty)}. \tag{3.3.34}$$

From Theorem (3.3.30) it can not only be inferred that $(2\omega)^{-1}\vec{\phi}_\infty$ is an element of the theta divisor $\Theta_{\vec{K}_\infty}$, i.e. the set of zeros of $\vartheta\left[\binom{1/2}{1/2}, \binom{0}{1/2}\right]$, but also that, in the case $g = 2$, $\Theta_{\vec{K}_\infty}$ is a manifold of complex dimension one. Note that the restriction to the theta divisor is only possible because ∞ is a branch point what is essential for the validity of theorem (3.3.30).

Since $\Theta_{\vec{K}_\infty}$ is a one-dimensional submanifold of $\mathbb{C}^2$, there is a one-to-one functional relation between the first and the second component of $(2\omega)^{-1}\vec{\phi}_\infty$. By the definition of $\vec{\phi}_\infty$ in (3.3.27) and by Eqs. (3.3.24) and (2.4.4) it can be inferred that

$$\vec{\phi}_\infty = \lim_{u_2 \to \infty} \vec{\phi} = \lim_{u_2 \to \infty} \vec{\varphi} - 2\int_{u_0}^{\infty} d\vec{z} = \int_{u_0}^{u_1} d\vec{z} - \int_{u_0}^{\infty} d\vec{z}. \tag{3.3.35}$$

The physical coordinate φ is given by (3.3.18),

$$\varphi = \int_{u_0}^{u_1} \frac{z\, dz}{\sqrt{P_{\text{SdS}}(z)}} + \varphi_0 = \int_{u_0}^{u_1} dz_2 + \varphi_0. \tag{3.3.36}$$

We insert this in (3.3.35) and obtain

$$\vec{\phi}_\infty = \begin{pmatrix} \int_{u_0}^{u_1} dz_1 - \int_{u_0}^{\infty} dz_1 \\ \varphi - \varphi_0 - \int_{u_0}^{\infty} dz_2 \end{pmatrix} = \begin{pmatrix} \int_{u_0}^{u_1} dz_1 - \int_{u_0}^{\infty} dz_1 \\ \varphi - \varphi_{in} \end{pmatrix}, \tag{3.3.37}$$

where $\varphi_{in} = \varphi_0 + \int_{u_0}^{\infty} dz_2$ depends only on the initial values u_0 and φ_0. Because $(2\omega)^{-1}\vec{\phi}_\infty$ is an element of the one-dimensional theta divisor $\Theta_{\vec{K}_\infty}$ there exists a function $f_{\vec{K}_\infty}$ such that $(2\omega)^{-1}\vec{\phi}_\infty = (2\omega)^{-1}\left(\begin{smallmatrix} f_{\vec{K}_\infty}(\varphi-\varphi_{in}) \\ \varphi - \varphi_{in} \end{smallmatrix}\right)$. Therefore, we finally obtain

$$r(\varphi) = \frac{M}{u(\varphi)} = -M\frac{\sigma_2(\vec{\phi}_\infty)}{\sigma_1(\vec{\phi}_\infty)} = -M\frac{\sigma_2}{\sigma_1}\left(\begin{pmatrix} f_{\vec{K}_\infty}(\varphi - \varphi_{in}) \\ \varphi - \varphi_{in} \end{pmatrix}\right), \tag{3.3.38}$$

where $\sigma(\vec{\phi}_\infty) = 0$.

This is the analytic solution of the equation of motion of a test particle in a Schwarzschild-(anti-)de Sitter space-time. This solution is valid in all regions of this space-time as well as for both signs of the cosmological constant, and can be computed with, in principle, arbitrary accuracy. The explicit computation of the solution is described in Appendix B.

3. Geodesics in spherically symmetric space-times

3.3.1.3 Periastron advance of bound timelike orbits

In the case that R_{SdS} has at least three positive zeros, we may have a bound orbit for some initial values. The periastron advance Δ_{peri} for such a bound orbit is given by the difference of the 2π-periodicity of the angle φ and the periodicity of the solution $r(\varphi)$ (which is the same as the periodicity of $u(\varphi) = \frac{M}{r(\varphi)}$ which corresponds to P_{SdS}, see (3.3.17)). Let us assume that the bound orbit corresponds to the interval $[u_1, u_2]$, where u_1 and u_2 are real and positive zeros of P_{SdS}, and that the path a_i of the homology basis of the Riemann surface of P_{SdS}, see Fig. 2.1, surrounds this real interval. Then the periastron advance is given by

$$\Delta_{\text{peri}} = 2\pi - 2\omega_{2i} = 2\pi - \oint_{a_i} \frac{u \, du}{\sqrt{P_{\text{SdS}}(u)}} = 2\pi - 2\int_{u_1}^{u_2} \frac{u \, du}{\sqrt{P_{\text{SdS}}(u)}}, \quad (3.3.39)$$

where $2\omega_{2i}$ is an element of the (canonically chosen) 2×4 matrix of periods $(2\omega, 2\omega')$ of the Riemann surface of $y^2 = P_{\text{SdS}}$, see Eq. (2.2.4). As Eq. (3.3.39) does not directly show the influence of the cosmological constant on the periastron advance, we now calculate the post-Schwarzschild limit of this expression in the case that the considered bound orbit is also bound in Schwarzschild space-time.

For doing so we first expand $u/\sqrt{P_{\text{SdS}}(u)}$ to first order in $\bar{\Lambda}$,

$$\frac{u}{\sqrt{P_{\text{SdS}}(u)}} \approx \frac{1}{\sqrt{P_S(u)}} - \frac{1}{2}\frac{u^2 + \mathcal{L}}{u^2 P_S(u)\sqrt{P_S(u)}}\bar{\Lambda}, \quad (3.3.40)$$

where P_S is the right hand side of Eq. (3.2.8), i.e. $P_S(u) = 2u^3 - u^2 + 2\mathcal{L}u + \mathcal{L}(E^2 - 1)$, which corresponds to the Schwarzschild case with $\bar{\Lambda} = 0$.

The next step is to integrate both terms involving P_S within the Weierstrass formalism, see for example [43]. Employing the substitution $u = 2y + \frac{1}{6}$ as in Sec. 3.2.1, P_S can be rewritten in the Weierstrass form

$$P_S(y) = 4(4y^3 - g_2 y - g_3) =: 4P_W(y), \quad (3.3.41)$$

where the Weierstrass invariants g_2, g_3 are given by (3.2.10),

$$g_2 = \frac{1}{12} - \epsilon\mathcal{L},$$
$$g_3 = \frac{1}{216} - \frac{1}{12}\epsilon\mathcal{L} - \frac{1}{4}\mathcal{L}(E^2 - \epsilon).$$

We assume that the orbit under consideration is bound not only in Schwarzschild-de Sitter but also in the corresponding Schwarzschild space-time. This means that the three largest real zeros $0 < u_1 < u_2 < u_3$ of P_{SdS} are positive and, thus, that the zeros $-\frac{1}{12} < y_1 < y_2 < y_3$ of P_W are all real. The square root $\sqrt{P_W}$ is branched over y_1, y_2 and y_3 and, thus, the elliptic function $\wp$ based on $\sqrt{P_W}$ has a purely real period 2ω and a purely imaginary period $2\omega'$. They are given by

$$2\omega = \oint_A \frac{dz}{\sqrt{P_W(z)}}, \quad 2\omega' = \oint_B \frac{dz}{\sqrt{P_W(z)}} \quad (3.3.42)$$

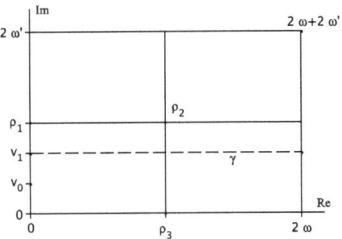

Figure 3.14: The fundamental rectangle R. The dashed line denotes the integration path γ.

where the path A runs around the branch cut from y_1 to y_2 counterclockwise and the path B around y_2 and y_3 clockwise. The branch of the square root in (3.3.42) is chosen such that $\sqrt{P_W}$ is negatively imaginary on (y_2, y_3). The branch points y_1, y_2, y_3 can be expressed in terms of the periods: $y_1 = \wp(\rho_1)$, $y_2 = \wp(\rho_2)$ and $y_3 = \wp(\rho_3)$ with $\rho_1 = \omega'$, $\rho_2 = \omega + \omega'$, and $\rho_3 = \omega$. The fundamental rectangle in the complex plane spanned by the periods $2\omega, 2\omega'$ of $\wp$ is denoted by $R = \{2x\omega + 2y\omega' \,|\, 0 \leq x, y < 1\}$, see Fig. 3.14.

For the canonical choice of the matrix of half-periods ω of $\sqrt{P_{\text{SdS}}}$, the integration path a_i in Eq. (3.3.39) corresponding to $[u_1, u_2]$ runs counterclockwise from u_1 to u_2 and back with conversed sign of the square root. Let the path γ be the preimage of a_i by $v \mapsto \wp(v) = y$ in the fundamental rectangle R. For a positive cosmological constant Λ, we have $u_1 < y_1 < y_2 < u_2$ and, thus, γ starts at some purely imaginary $\gamma(0) = v_1 \in R$ with $0 < \operatorname{Im}(v_1) \leq \operatorname{Im}(\rho_1)$ and goes straight to $\gamma(1) = v_1 + 2\omega$. Then, for any rational function f, we obtain

$$\oint_{a_i} \frac{f(z)\,dz}{\sqrt{P_W(z)}} = \int_\gamma f(\wp(v))dv \,. \tag{3.3.43}$$

This is derived from the differential equation (2.1.2)

$$\wp'(v) = \sqrt{4\wp(v)^3 - g_2\wp(v) - g_3} = \sqrt{P_W(\wp(v))}\,,$$

where the branch of the square root was chosen to be consistent with the sign of $\wp'$.

The integration of the first part on the right-hand side of (3.3.40) is straightforward and yields the Schwarzschild period

$$\oint_{a_i} \frac{du}{\sqrt{P_S(u)}} = \oint_{a_i} \frac{2dy}{\sqrt{4P_W(y)}} = \int_{v_1}^{v_1+2\omega} dv = 2\omega \,. \tag{3.3.44}$$

The integration of the second part on the right-hand side of (3.3.40) is more involved and can be

3. Geodesics in spherically symmetric space-times

carried out along the lines of Thm. 2.5. First, we substitute $y = \wp(v)$ obtaining

$$\oint_{a_i} \frac{u^2 + \mathcal{L}}{u^2 P_S(u)\sqrt{P_S(u)}} du = \oint_A \frac{(2y + \frac{1}{6})^2 + \mathcal{L}}{(2y + \frac{1}{6})^2 \cdot 4P_W(y)\sqrt{4P_W(y)}} 2dy$$

$$= \frac{1}{4} \int_{v_1}^{v_1+\omega} F_1(v) + \mathcal{L} F_2(v)\, dv, \qquad (3.3.45)$$

where

$$F_1(v) = \frac{1}{P_W(\wp(v))} = \frac{1}{(\wp'(v))^2},$$

$$F_2(v) = \frac{1}{(2\wp(v) + \frac{1}{6})^2 P_W(\wp(v))} = \frac{1}{(2\wp(v) + \frac{1}{6})^2 (\wp'(v))^2}.$$

Second, we have to determine the poles p_i with multiplicity m_i and the corresponding constants A_i^j, $j = 0, \ldots, m_i - 1$, see Eq. (2.1.20), for F_1 and F_2. Finally, an integration (2.1.21) can be carried through. The details of this procedure can be found in appendix A. As a result we obtain the first order approximation of the periastron advance with respect to $\bar{\Lambda}$:

$$\Delta_{\text{peri}} = 2\pi - \oint_{a_i} \frac{u\, du}{\sqrt{P_{\text{SdS}}(u)}}$$

$$= 2\pi - 2\omega - \frac{\bar{\Lambda}}{8} \left\{ \sum_{j=1}^{3} \frac{-\eta_1 - y_j \omega}{\wp''(\rho_j)^2} \left(1 + \frac{\mathcal{L}}{(2y_j + \frac{1}{6})^2}\right) \right.$$

$$\left. + \mathcal{L} \left[\frac{\omega - 12\eta_1}{24\wp'(v_0)^4} + \frac{3}{2} \frac{\wp''(v_0)}{\wp'(v_0)^5} (\eta_1 v_0 + \omega \zeta(v_0)) \right] \right\} + \mathcal{O}(\bar{\Lambda}^2), \qquad (3.3.46)$$

where v_0 is such that $\wp(v_0) = -\frac{1}{12}$, y_j are the zeros of $P_W(y)$, $\wp(\rho_j) = y_j$, and η_1 is the real period of second kind.

The terms in this expression involving ρ_j and v_0 can partly be replaced by terms containing the Weierstrass invariants g_2 and g_3. From the differential equation (2.1.2) we derive

$$\wp'(v_0) = \sqrt{4\wp(v_0)^3 - g_2 \wp(v_0) - g_3} = \sqrt{-\frac{1}{432} + \frac{g_2}{12} - g_3}. \qquad (3.3.47)$$

The first derivative of (2.1.2) yields $2\wp'' = 12\wp^2 - g_2$ and, thus, gives

$$\wp''(\rho_j) = 6y_j^2 - \frac{1}{2}g_2 \quad \text{and} \quad \wp''(v_0) = \frac{1}{24} - \frac{1}{2}g_2, \qquad (3.3.48)$$

where the g_2, g_3, as well as the zeros of P_W can be expressed in terms of E^2 and $\mathcal{L}$.

The result (3.3.46) gives the post-Schwarzschild periastron advance in a closed algebraic form. The advantage of this result is that no further integration is needed. Another advantage lies in the fact that only elliptic functions and related quantities are used, which are well described and tabulated in mathematical books and which are also well implemented in common commercial math programs.

What is still left to do is to express the result (3.3.46) in terms of, e.g., $r_{\min}$ and $r_{\max}$ or, equivalently, in terms of the semimajor axis and the eccentricity. These quantities can be directly observed and also have the advantage that an expansion in terms of $M/r_{\min}$ and $M/r_{\max}$ can be performed giving in addition a post-Newtonian expression.

Let us apply these formulas to the perihelion advance of Mercury and compare with the results of Kraniotis and Whitehouse, [40]. We take the values indicated in Tab. 3.4 for the Schwarzschild-radius $2M$, the angular momentum per unit mass L_M, and the energy per unit mass E_M given in [40]. These values lead to the zeros and periods in Tab. 3.4, which all compare well to the results in [40]. Also the physical data, i.e. the aphel r_a, the perihel r_p and the perihelion advance in Schwarzschild-space-time Δ^S, compare well to [40] and also to observations[1]. Here we used the rotation period 87.97 days of Mercury and 100 SI-years per century to determine the unit $\mathrm{arcsec\,cy^{-1}}$.

The first order post-Schwarzschild correction $\Delta_{\mathrm{corr}}^{\mathrm{SdS}}$ to the perihelion advance can now be calculated from formula (3.3.46). For a cosmological constant of $\Lambda = 10^{-51}\mathrm{m}^{-2}$ we obtain for the parameters which appear in the expansion (3.3.46) and the correction $\Delta_{\mathrm{corr}}^{\mathrm{SdS}}$ the values indicated in Tab. 3.4,

$$\Delta_{\mathrm{corr}}^{\mathrm{SdS}} = 5.82 \cdot 10^{-17} \,\mathrm{arcsec\,cy^{-1}}. \tag{3.3.49}$$

This result also compares well to [40] where the perihel advance of Mercury does not change within the given accuracy when considered in Schwarzschild-de Sitter space-time. The value of the correction is also far beyond the measurement accuracy of $0.002\,\mathrm{arcsec\,cy^{-1}}$ for the perihelion advance of Mercury.

However, for an extreme case the influence of the cosmological constant on the periastron advance may become measurable. The orbital data of quasar QJ287 reported in [66, 67] indicates that the correction to the periastron advance $\Delta_{\mathrm{corr}}^{\mathrm{SdS}}$ will be some orders of magnitude larger than the correction in the case of Mercury. Indeed, when we calculate from this data the energy E and the angular momentum parameter $\mathcal{L}$,

$$E^2 = 0.982166, \quad \mathcal{L} = 0.092317, \tag{3.3.50}$$

we obtain

$$\Delta_{\mathrm{corr}}^{\mathrm{SdS}} \approx 10^{-13} \,\mathrm{arcsec\,cy^{-1}}. \tag{3.3.51}$$

3.3.1.4 On the Pioneer anomaly

The term *Pioneer anomaly* describes an anomalous constant acceleration of $a_{\mathrm{Pioneer}} = (8.47 \pm 1.33) \times 10^{-10}\,\mathrm{m/s^2}$ towards the inner solar system found in the orbital data of the Pioneer 10 and 11 spacecraft, which were the first to enter the outer solar system. In 2002, Anderson et al. checked a number

---
[1] http://history.nasa.gov/SP-423/intro.htm

3. Geodesics in spherically symmetric space-times

parameters	zeros of P_W
$2M = 2953.25008$ m	-0.083333322758379
$\frac{2M}{L_M^2} = 1.1849627128268641 \times 10^{-24}\,\text{m}^2/\text{s}^2$	-0.083333317283501
$\sqrt{E_M} = 0.029979245417779875 \times 10^{10}$ m/s	0.166666640041880

periods	results
$\omega = 3.1415929045225246$	$r_\text{a} = 6.981708938652731 \times 10^{10}$m
$\omega' = 20.4093916393851799 i$	$r_\text{p} = 4.600126052898539 \times 10^{10}$m
$\tau = 6.4965106109084187 i$	$\Delta^\text{S} = 42.980165$ arcsec cy^{-1}

(a) Predicted perihelion shift Δ^S and peri- and apoapsis r_p, r_a of Mercury in Schwarzschild space-time.

values in Eq. (3.3.46)	Correction due to $\bar{\Lambda}$
$\eta_1 = -0.2617993700131308$	
$\wp'^2(v_0) = -1.6972622347915708 \times 10^{-16}$	$\Delta^\text{SdS}_\text{corr} = 5.82 \times 10^{-17}$ arcsec cy^{-1}
$\wp''(v_0) = 1.3312392184539657 \times 10^{-8}$	
$\zeta(v_0) = 1.0121093196146584\, i$	

(b) Predicted first order correction $\Delta^\text{SdS}_\text{corr}$ due to the cosmological constant $\Lambda = 10^{-51}\text{m}^{-2}$.

Table 3.4: Predicted perihelion shift of Mercury in Schwarzschild and Schwarzschild-de Sitter space-time

of possible reasons for this anomalous constant acceleration [26], but were not able to give a convincing explanation. Although at least a part of the Pioneer anomaly is caused by thermal effects [68, 69], this explanation is inconsistent with the constancy of the acceleration over time. However, a connection of the Pioneer anomaly with the cosmological constant can be suspected inspired by the numerical coincidence that a_Pioneer is of the same order as cH, where c is the speed of light and H the Hubble constant. For a definite check of the effect of a nonvanishing cosmological constant on the spacecraft, we apply the obtained analytical solution in order to decide whether an observable influence on the Pioneer satellites arises. From [70] we deduce the energy and angular momentum of the Pioneers after their last flybys at Jupiter and Saturn, respectively, with respect to the barycenter of the inner solar system, i.e. the Sun, Mercury, Venus and Earth-Moon. This means that we used the value

$$2M = \frac{2GM}{c^2} = 2.95326676236345\,\text{km}\,, \qquad (3.3.52)$$

3.3. S- and RN-(anti-)dS space-times

for the Schwarzschild radius, derived from $GM = 1.00000565\, k^2 (\text{AU}^3/\text{day}^2)$ with Gauss' constant $k = 0.01720209895$ defining the astronomical unit AU. Here all numbers are taken with 12 digits what corresponds to the today's accuracy of solar system ephemerides.

In the case of Pioneer 10, the velocity at infinity $v_\infty = 11.322\,\text{km s}^{-1}$ taken from [70] gives us the energy per unit mass $E_M = c^2 + \frac{1}{2}v_\infty^2$ and therefore the energy parameter,

$$E^2 = \frac{E_M^2}{c^4} = 1.00000000143. \qquad (3.3.53)$$

The angular momentum per unit mass is given by $L_M = qv$, where v is the velocity at periapsis distance q defined by $v = \sqrt{2GM\left(\frac{1}{q} + \frac{1}{2a}\right)}$ and a is the semimajor axis. From this we derive the parameter $\mathcal{L}$,

$$4\mathcal{L} = \frac{(2M)^2 c^2}{L_M^2} = 2.85557237382 \times 10^{-9}. \qquad (3.3.54)$$

In the case of Pioneer 11 we obtain for the parameters E^2 and $\mathcal{L}$

$$\begin{aligned} E^2 &= 1.00000000122, \\ 4\mathcal{L} &= 1.34074057459 \times 10^{-9}. \end{aligned} \qquad (3.3.55)$$

With these coefficients, now we are able to determine the exact orbits of Pioneer 10 and 11 in the cases $\Lambda = 0$ and $\Lambda = 10^{-45}\,\text{km}^{-2}$. From these exact orbits we calculate the differences in position (in m) for a given angle φ (in rad) and the difference in the angle (in rad) for a given distance r (in m) of a test particle moving in a space-time with and without the cosmological constant. The Pioneer anomaly appeared in a heliocentric distance of about 20 to 70 AU. For r in this range, we compute now the difference $\varphi_{\Lambda=0}(r) - \varphi_{\Lambda\neq 0}(r)$ in azimuthal position with and without cosmological constant for both spacecraft. Regarding Pioneer 10, the difference is in the scale of 10^{-19}rad, which corresponds to an azimuthal difference in position of about 10^{-6}m. For Pioneer 11, the difference is in the scale of 10^{-18}rad, which corresponds to an azimuthal difference in position of about 10^{-5}m.

The range of 20 to 70 AU corresponds to an angle between 0.4π and 0.6π if $\varphi_0 = 0$ corresponds to the periapsis. In this range, we compute the radial difference $r_{\Lambda=0}(\varphi) - r_{\Lambda\neq 0}(\varphi)$ also for both spacecraft. For Pioneer 10 we obtain a difference in the scale of 10^{-5}m, for Pioneer 11 in the scale of 10^{-4}m.

Therefore, we can state that for the present value of the cosmological constant the forms of the Pioneer 10 and 11 orbits practically do not change. Although for a definite estimate of the differences of the Pioneer orbits in Schwarzschild and Schwarzschild-de Sitter space-time also the time course of these orbits has to be analyzed, the time variable is influenced by the cosmological constant in the same way as the radial coordinate and no change in our statement will occur. Therefore, the influence of the cosmological constant on the orbits cannot be held responsible for the observed anomalous acceleration of the Pioneer spacecraft.

3. Geodesics in spherically symmetric space-times

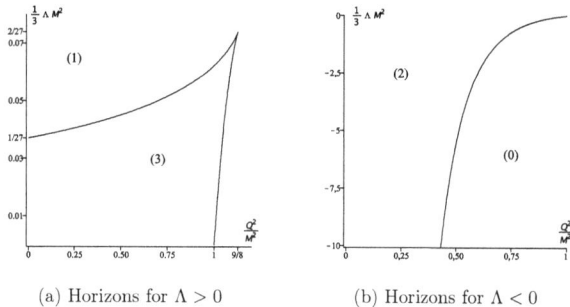

(a) Horizons for $\Lambda > 0$ (b) Horizons for $\Lambda < 0$

Figure 3.15: Number of horizons in Reissner-Nordström-(anti-)de Sitter space-times dependent on Λ and Q^2. In (a), there are three horizons in the region bounded by the black line, (a cosmological, an event, and a Cauchy horizon) and one outside. In (b), there are two horizons above the black line (an event and a Cauchy horizon) and none below.

3.3.2 Geodesics in Reissner-Nordström-(anti-)de Sitter space-times

3.3.2.1 Types of orbits

Now the most general space-time with the most complicated types of orbits in this chapter is treated. Analogous to the foregoing sections, all types of orbits can be determined from the right hand side of Eq. (3.1.3)

$$\left(\frac{dr}{d\varphi}\right)^2 = R(r) = \frac{r^4}{L^2}\left(E^2 - \frac{\Delta_{\text{RNdS}}}{r^2}\left(\epsilon + \frac{L^2}{r^2}\right)\right), \quad (3.1.3)$$

where $\Delta_{\text{RNdS}} = r^2 - 2Mr - \frac{1}{3}\Lambda r^4 + Q^2$. It is convenient to introduce the dimensionless quantities

$$\bar{r} := \frac{r}{M}, \quad \mathcal{L} := \frac{M^2}{L^2}, \quad \bar{Q} := \frac{Q}{M}, \quad \bar{\Lambda} := \frac{1}{3}\Lambda M^2, \quad (3.3.56)$$

into Eq. (3.1.3) giving

$$\left(\frac{d\bar{r}}{d\varphi}\right)^2 = \epsilon\bar{\Lambda}\mathcal{L}\bar{r}^6 + ((E^2 - \epsilon)\mathcal{L} + \bar{\Lambda})\bar{r}^4 + 2\epsilon\mathcal{L}\bar{r}^3 - (1 + \epsilon\bar{Q}^2\mathcal{L})\bar{r}^2 + 2\bar{r} - \bar{Q}^2$$

$$=: R_{\text{RNdS}}(\bar{r}). \quad (3.3.57)$$

In Reissner-Nordström-(anti-)de Sitter space-time the horizons are located at the positive real zeros of Δ_{RNdS}. By Descartes' rule of signs it can be inferred that for $\Lambda > 0$ there are up to three

3.3. S- and RN-(anti-)dS space-times

horizons whereas for $\Lambda < 0$ there are up to two horizons. One of the horizons for $\Lambda > 0$ is cosmological and appears in Schwarzschild-de Sitter space-time, too. The other two, if present, are approximately located at $M \pm \sqrt{M^2 - Q^2}$, perturbed by the small cosmological constant Λ. In the following it is assumed that the cosmological constant $\bar{\Lambda}$ and the electric charge $\bar{Q}^2$ are chosen such that the space-time has no naked singularity. In Fig. 3.15 the number of horizons dependent on $\bar{\Lambda}$ and $\bar{Q}^2$ is shown.

Eq. (3.3.57) implies that $R_\mathrm{RNdS} \geq 0$ is a necessary condition for the existence of a geodesic. Therefore, the real and positive zeros of R_RNdS determine the type of motion. For varying parameters E^2 and $\mathcal{L}$, the number of positive real zeros of R_RNdS changes only if multiple zeros occur. This happens at $\bar{r} = x$ iff

$$R_\mathrm{RNdS}(\bar{r}) = (\bar{r} - x)^2 (a_4 \bar{r}^4 + a_3 \bar{r}^3 + a_2 \bar{r}^2 a_1 \bar{r} + a_0) \tag{3.3.58}$$

for some constant a_i. The resulting 7 equations can be solved for E^2 and $\mathcal{L}$ dependent on ϵ and the parameters of the black hole $\bar{\Lambda}$ and $\bar{Q}$ by

$$E^2(x) = 2 \frac{(x(x-2) + \bar{Q}^2 - \bar{\Lambda} x^4)^2}{x^2 (x^2 - 3x - 2\bar{Q}^2)}, \quad \mathcal{L}(x) = \frac{x^2 - 3x - 2\bar{Q}^2}{(x - \bar{Q}^2 - \bar{\Lambda} x^4) x^2} \tag{3.3.59}$$

for $\epsilon = 1$, where x is the position of the double zero, and by

$$\mathcal{L} = \frac{2(1 + \sqrt{9 - 8\bar{Q}^2})}{E^2 (3 + \sqrt{9 - 8\bar{Q}^2})^3} - \frac{\bar{\Lambda}}{E^2} \tag{3.3.60}$$

for $\epsilon = 0$. As in Reissner-Nordström space-time it is obvious from Eq. (3.3.60) that for $\bar{Q}^2 > \frac{9}{8}$ and $\epsilon = 0$ there are no double zeros.

In Figs. 3.16 and 3.17 regions of different types of geodesic motion are shown for varying values of $\bar{Q}^2, \bar{\Lambda} > 0$ and ϵ. As in Schwarzschild-de Sitter space-time 3 different regions can be found:

(I) $R_\mathrm{RNdS}(\bar{r})$ has 3 positive real zeros $r_1 < r_2 < r_3$ with $R_\mathrm{RNdS}(\bar{r}) \geq 0$ for $r_1 \leq \bar{r} \leq r_2$ and $r_3 \leq \bar{r}$. Possible orbit types: flyby and bound orbits.

(II) $R_\mathrm{RNdS}(\bar{r})$ has 1 positive real zero r_1 and $R_\mathrm{RNdS}(\bar{r}) \geq 0$ for $\bar{r} \geq r_1$. Possible orbit types: flyby orbits.

(IV) $R_\mathrm{RNdS}(\bar{r})$ has 5 positive real zeros $r_i < r_{i+1}$ with $R_\mathrm{RNdS}(\bar{r}) \geq 0$ for $r_1 \leq \bar{r} \leq r_2, r_3 \leq \bar{r} \leq r_4$, and $r_5 \leq \bar{r}$. Possible orbit types: flyby and two different bound orbits.

As in Reissner-Nordström space-time, for light ($\epsilon = 0$) only regions (I) and (II) appear. If additionally $\bar{Q}^2 > \frac{9}{8}$ also region (I) vanishes and only region (II) is possible. In Fig. 3.18 sample effective potentials are shown for each of the regions above.

3. Geodesics in spherically symmetric space-times

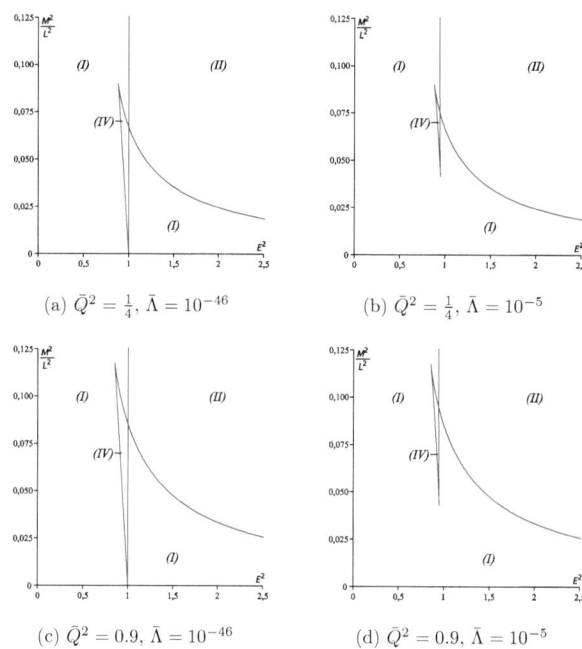

Figure 3.16: Regions of different types of timelike geodesics in Reissner-Nordström-de Sitter space-time ($\epsilon = 1$) for different values of $\bar{Q}^2$ and $\bar{\Lambda}$.

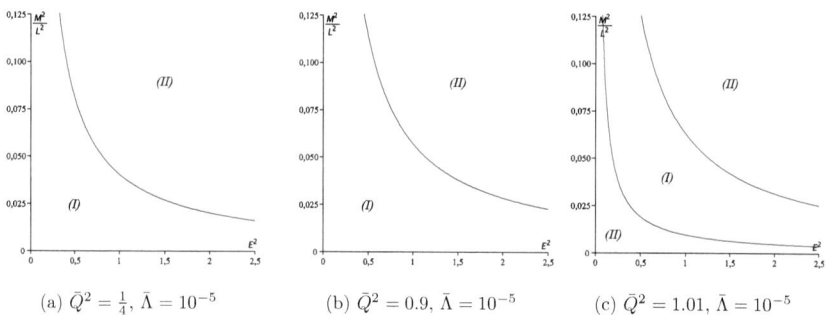

Figure 3.17: Regions of different types of null geodesics in Reissner-Nordström-de Sitter space-time ($\epsilon = 0$) for different values of $\bar{Q}^2$ and $\bar{\Lambda}$. For a naked singularity with $1 < \bar{Q}^2 < \frac{9}{8}$ a second region (II) appears which grows with $\bar{Q}^2$ until region (I) vanishes at $\bar{Q}^2 = \frac{9}{8}$. For effective potentials see Fig. 3.18.

3.3. S- and RN-(anti-)dS space-times

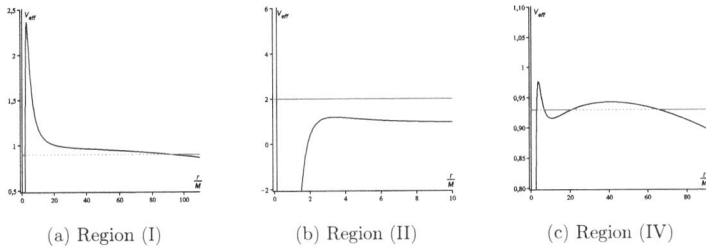

(a) Region (I) (b) Region (II) (c) Region (IV)

Figure 3.18: Effective potentials for different regions of geodesic motion in Reissner-Nordström-de Sitter space-time. The horizontal lines denote E^2.

region	pos. zeros	range of $\bar{r}$	types of orbits
I	3		flyby, bound
II	1		flyby
IV	5		flyby, 2x bound

Table 3.5: Orbit types in Reissner-Nordström-de Sitter space-time. The second column gives the number of positive zeros of the polynomial R_{RNdS}. In the third column, the thick lines represent the range of orbits and turning points are shown by thick dots. The small vertical line denotes $r = 0$.

The geodesics in Reissner-Nordström-(anti-)de Sitter space-times compare to those for $Q^2 = 0$ in the same way as for a vanishing cosmological constant, i.e. in each region appears an additional positive real zero, which prevents test particles and light from falling into the singularity at $\bar{r} = 0$. Again, many-world bound orbits and two-world escape orbits are possible. Compared to Reissner-Nordström space-time, the nonvanishing cosmological constant causes the same effects as in Schwarzschild-de Sitter space-time described in Sec. 3.3, in particular orbits may reach infinity for $E^2 < 1$ and some flyby orbits are reflected at the potential barrier induced by Λ. However, the two different bound orbits in region (IV), one of which is a many-world orbit, have different periods in contrast to Reissner-Nordström space-time. In Fig. 3.19 an example for a timelike geodesic in each of the regions (I), (II), and (IV) is shown. A schematic overview of possible orbits types in the different regions can be found in Tab. 3.5.

As for the other space-times considered in this book, exceptional orbits correspond to multiple zeros of R_{RNdS} and are located at the boundaries of regions (I), (II), and (IV). The associated parameters E^2 and $\mathcal{L}$ are given by (3.3.59) for test particles and by (3.3.60) for light. A substitution of (3.3.59)

59

3. Geodesics in spherically symmetric space-times

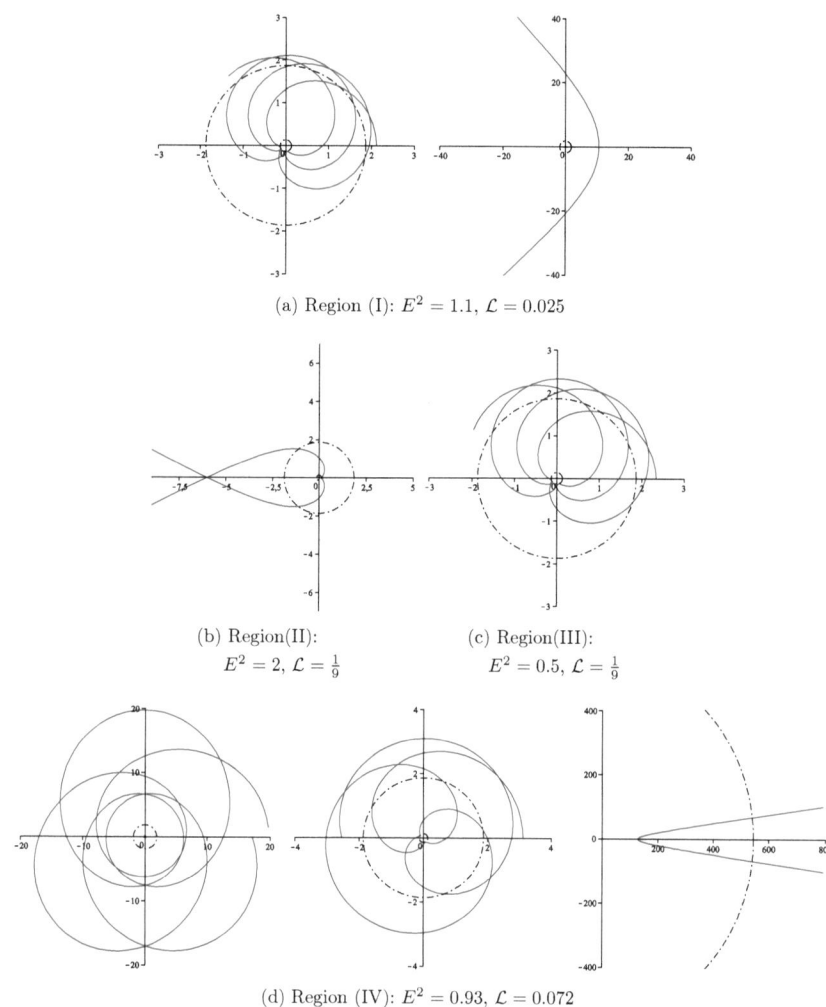

(a) Region (I): $E^2 = 1.1$, $\mathcal{L} = 0.025$

(b) Region(II):
$E^2 = 2$, $\mathcal{L} = \frac{1}{9}$

(c) Region(III):
$E^2 = 0.5$, $\mathcal{L} = \frac{1}{9}$

(d) Region (IV): $E^2 = 0.93$, $\mathcal{L} = 0.072$

Figure 3.19: Timelike geodesics in Reissner-Nordström-de Sitter space-time with $\bar{Q}^2 = \frac{1}{4}$ for every region of orbit types. The plots are in units of M and dashed lines mark the horizons. All bound orbits with the exception of the bound orbit on the left in (d) crosses the inner Cauchy horizon just barely and, thus, are many-world bound orbits. Also, the flyby orbit in (b) is a two-world orbit.

into $\frac{d^2}{d\bar{r}^2}R_{\mathrm{RNdS}}(\bar{r})$ yields

$$\frac{d^2}{d\bar{r}^2}R_{\mathrm{RNdS}}(x) = -2\frac{x^3 - 6x^2 - \bar{\Lambda}x^5(4x-15) - 12x^4\bar{\Lambda}\bar{Q}^2 + 9\bar{Q}^2 x - 4\bar{Q}^4}{x^2(x - \bar{Q}^2 - \bar{\Lambda}x^4)}. \tag{3.3.61}$$

For positive $\bar{\Lambda}$ all real zeros of this expression are also positive by Descartes' rule of signs, but for small $\bar{\Lambda}$ only 2 zeros $\bar{r}_{tr,1}$, $\bar{r}_{tr,2}$ are associated with finite values $E^2 \geq 0$ and $\mathcal{L} \geq 0$. The smaller of these two triple zeros $\bar{r}_{tr,1}$ is located near $x = 6$ for small $\bar{\Lambda}$ and $\bar{Q}^2$ and the corresponding pair $(E^2, \mathcal{L})$ is given by the upper corner in the boundary of region (IV), whereas the other, lower corner is associated with the larger triple zero $\bar{r}_{tr,2}$. The triple zero $\bar{r}_{tr,1}$ is a saddle point and the radial coordinate of the innermost stable circular orbit. All double zeros x with $0 < x < \bar{r}_{tr,1}$ or $x > \bar{r}_{tr,2}$ are minima, i.e. radial coordinates of unstable circular orbits, but x with $\bar{r}_{tr,1} < x < \bar{r}_{tr,2}$ are stable circular orbits. Thus, as in Schwarzschild-de Sitter space-time $\bar{r}_{tr,2}$ can be called the outermost stable circular orbit. For light the situation is the same as in Reissner-Nordström space-time, see Sec. 3.2.2: a double zero can only exist if $\bar{Q}^2 \leq \frac{9}{8}$ and is then always located at $\bar{r} = \frac{3}{2} + \frac{1}{2}\sqrt{9 - 8\bar{Q}^2}$, which is a saddle point for a naked singularity and $\bar{Q}^2 = \frac{9}{8}$, and else a minimum corresponding to unstable circular orbits.

3.3.2.2 Analytical solution of geodesic equations

Like in Schwarzschild-(anti-)de Sitter space-time, the geodesic equation (3.1.3) for the most general of space-times considered in this chapter, the Reissner-Nordström-(anti-)de Sitter space-time,

$$\left(\frac{dr}{d\varphi}\right)^2 = R_{\mathrm{RNdS}}(r) = \frac{r^4}{L^2}\left(E^2 - \frac{\Delta_{\mathrm{RNdS}}}{r^2}\left(\epsilon + \frac{L^2}{r^2}\right)\right), \tag{3.1.3}$$

where $\Delta_{\mathrm{RNdS}} = r^2 - 2Mr - \frac{1}{3}\Lambda r^4 + Q^2$, is in general of hyperelliptic type. Consequently, the analytical solution of this equation can be found in the same way as the solution of the geodesic equation in Schwarzschild-(anti-)de Sitter space-time presented in Sec. 3.3.

With the introduction of a new variable $u = \frac{M}{r}$ analogous to the foregoing sections Eq. (3.3.57) is transformed to

$$\left(\frac{du}{d\varphi}\right)^2 = -\bar{Q}^2 u^4 + 2u^3 - (1 + \epsilon\bar{Q}^2\mathcal{L})u^2 + 2\epsilon\mathcal{L}u + (\mathcal{L}(E^2 - \epsilon) + \bar{\Lambda}) + \epsilon\bar{\Lambda}\mathcal{L}\frac{1}{u^2} \tag{3.3.62}$$

with the dimensionless parameters introduced in (3.3.56)

$$\mathcal{L} = \frac{M^2}{L^2}, \quad \bar{Q} = \frac{Q}{M}, \quad \bar{\Lambda} = \frac{1}{3}\Lambda M^2.$$

For $\epsilon = 1$ Eq. (3.3.62) should be rewritten as

$$\left(u\frac{du}{d\varphi}\right)^2 = P_{\mathrm{RNdS}}(u) \tag{3.3.63}$$

3. Geodesics in spherically symmetric space-times

where

$$P_{\text{RNdS}}(u) := -\bar{Q}^2 u^6 + 2u^5 - (1 + \bar{Q}^2 \mathcal{L}) u^4 + 2\mathcal{L} u^3 + \left(\mathcal{L}(E^2 - 1) + \bar{\Lambda}\right) u^2 + \mathcal{L}\bar{\Lambda}. \tag{3.3.64}$$

However, analogous to the situation in Reissner-Nordström space-time in Sec. 3.2.2, the substitution

$$r = \pm \frac{1}{\xi} + r_{\text{RNdS}}, \tag{3.3.65}$$

where r_{RNdS} is the zero of $R_{\text{RNdS}}(r)$, is more convenient as it reduces Eq. (3.3.57) such that the resulting equation is of type (3.3.63) but with a polynomial of degree 5 if $\epsilon = 1$ and of degree 3 if $\epsilon = 0$ on the right hand side. For $\epsilon = 0$ we can choose the positive sign in (3.3.65) and the resulting equation reads

$$\left(\frac{d\xi}{d\varphi}\right)^2 = \sum_{j=0}^{3} b_j \xi^j, \quad b_j = \frac{1}{(4-j)!} \frac{d^{(4-j)} R}{dr^{4-j}}(r_{\text{RNdS}}), \tag{3.3.66}$$

but for $\epsilon = 1$

$$\left(\xi \frac{d\xi}{d\varphi}\right)^2 = \sum_{j=0}^{5} a_j \xi^j, \quad a_j = \frac{(-1)^j}{(6-j)!} \frac{d^{(6-j)} R}{dr^{6-j}}(r_{\text{RNdS}}). \tag{3.3.67}$$

Here the sign of the substitution (3.3.65) should be chosen such that the leading coefficient a_5 is positive.

Null geodesics For $\epsilon = 0$ Eq. (3.3.66) is of elliptic type and can be solved analogously to the geodesic equation in Schwarzschild space-time,

$$r(\varphi) = \frac{b_3}{4\wp(\varphi - \varphi_{in}) - \frac{b_2}{3}} + r_{\text{RNdS}}, \tag{3.3.68}$$

where

$$\varphi_{in} = \varphi_0 + \int_{y_0}^{\infty} \frac{dz}{\sqrt{4y^3 - g_2 - g_3}}, \quad y_0 = \frac{1}{4}\left(\frac{\pm b_3}{r_0 - r_{\text{RNdS}}} + \frac{b_2}{3}\right) \tag{3.3.69}$$

with the standard expressions (2.1.11) of g_2 and g_3

$$g_2 = \frac{1}{16}\left(\frac{4}{3} b_2^2 - 4 b_1 b_3\right),$$

$$g_3 = \frac{1}{16}\left(\frac{1}{3} b_1 b_2 b_3 - \frac{2}{27} b_2^3 - b_0 b_3^2\right).$$

3.4. Higher-dimensional space-times

Timelike geodesics For $\epsilon = 1$ Eq. (3.3.67) is of hyperelliptic type as the geodesic equation in Schwarzschild-de Sitter space-time (3.3.17) and can be solved analogously to Sec. 3.3. The analytical solution is

$$r(\varphi) = \pm \frac{1}{\xi(\varphi)} + r_{\text{RNdS}} = \mp \frac{\sigma_2}{\sigma_1} \left(\begin{pmatrix} f_{\bar{K}_\infty}(\varphi - \varphi_{in}) \\ \varphi - \varphi_{in} \end{pmatrix} \right) + r_{\text{RNdS}}, \qquad (3.3.70)$$

where $\sigma \left(\begin{pmatrix} f_{\bar{K}_\infty}(x) \\ x \end{pmatrix}; \tau \right) = 0$,

i.e. $f_{\bar{K}_\infty}$ is the function that describes the one-dimensional theta divisor $\Theta_{\bar{K}_\infty}$, cf. (3.3.38). Note that the function $f_{\bar{K}_\infty}$ differs to the function used in (3.3.38) in the sense that it corresponds to a different normalized period matrix τ.

3.4 Higher-dimensional space-times

In this section it will be shown that it is also possible to apply the method for solving geodesic equations in terms of hyperelliptic functions presented in Sec. 3.3 to the geodesic equations in higher-dimensional static and spherically symmetric space-times. As an example, the geodesic equations in up to 7-dimensional Schwarzschild space-times are solved in the following. However, possible types of orbits are only discussed for a 6-dimensional Schwarzschild space-time as we only want to show the applicability of the method presented in Sec. 3.3. The solution of geodesic equations in more general static and spherically symmetric higher-dimensional space-times together with a complete discussion of orbits can be found in [53].

The metric of a Schwarzschild space-time in d dimensions is given by [56]

$$ds^2 = \left(1 - \left(\frac{2M}{r}\right)^{d-3}\right) dt^2 - \left(1 - \left(\frac{2M}{r}\right)^{d-3}\right)^{-1} dr^2 - r^2 d\Omega_{d-2}^2, \qquad (3.4.1)$$

where $d\Omega_1^2 = d\varphi^2$ and $d\Omega_{i+1}^2 = d\theta_i + \sin^2\theta_i d\Omega_i^2$ for $i \geq 1$. Because of the spherical symmetry, we again restrict the considerations to the equatorial plane by setting $\theta_i = \frac{\pi}{2}$ for all i. With the normalization condition $g_{\mu\nu} \frac{dx^\mu}{ds} \frac{dx^\nu}{ds} = \epsilon$, the conserved energy E, and the angular momentum L as well as the substitution $\bar{r} = \frac{r}{M}$ the geodesic equation reduces to

$$\left(\frac{d\bar{r}}{d\varphi}\right)^2 = \mathcal{L}(E^2 - \epsilon)\bar{r}^4 - \bar{r}^2 + 2^{d-3}\epsilon\mathcal{L}\bar{r}^{7-d} + 2^{d-3}\bar{r}^{5-d} = R_d(\bar{r}), \qquad (3.4.2)$$

where the parameters $\mathcal{L} = \frac{M^2}{L^2}$ and E^2 have the same meaning as in the 4-dimensional Schwarzschild case (3.2.1). For $d = 4$ this equation reduces of course to the Schwarzschild case [6]. For $d = 5$ a

3. Geodesics in spherically symmetric space-times

substitution $\bar{r} = x^2$ reduces the differential equation (3.4.2) to

$$\left(\frac{dx}{d\varphi}\right)^2 = \mathcal{L}(E^2 - \epsilon)x^2 + (4\epsilon\mathcal{L} - 1)x + 4, \tag{3.4.3}$$

which can be solved in terms of elementary functions, see [31].

In the case of a 6-dimensional Schwarzschild space-time, however, the differential equation (3.4.2) can be rewritten as

$$\left(\bar{r}\frac{d\bar{r}}{d\varphi}\right)^2 = \mathcal{L}(E^2 - \epsilon)\bar{r}^6 - \bar{r}^4 + 2^{d-3}\epsilon\mathcal{L}\bar{r}^3 + 2^{d-3}\bar{r}, \tag{3.4.4}$$

or, with the substitution $u = \frac{1}{\bar{r}}$

$$\left(\frac{du}{d\varphi}\right)^2 = 8u^5 + 8\epsilon\mathcal{L}u^3 - u^2 + \mathcal{L}(E^2 - \epsilon). \tag{3.4.5}$$

This now can be solved in terms of hyperelliptic functions with the method used for the geodesic equation in Schwarzschild-(anti-)de Sitter space-time. The only difference is that the physical angle φ is now given by

$$\varphi - \varphi_0 = \int_{u_0}^{u} \frac{du'}{\sqrt{P_5(u')}} = \int_{u_0}^{u} dz_1 \tag{3.4.6}$$

what corresponds to the holomorphic differential dz_1 rather than to dz_2 as it was in the Schwarzschild-de Sitter case, see (2.2.2) and (3.3.36). This means that the solution of the geodesic equation in 6-dimensional Schwarzschild space-time is given by

$$r(\varphi) = \frac{M}{u(\varphi)} = -M\frac{\sigma_2(\vec{\varphi}_{\Theta,6})}{\sigma_1(\vec{\varphi}_{\Theta,6})} = -M\frac{\sigma_2}{\sigma_1}\left(\begin{pmatrix}\varphi - \varphi_{\text{in}} \\ f_{\vec{K}_\infty}(\varphi - \varphi_{\text{in}})\end{pmatrix}\right), \tag{3.4.7}$$

where $\sigma(\vec{\varphi}_{\Theta,6}; \tau) = 0$.

Here the function $f_{\vec{K}_\infty}$ describes the theta divisor $\Theta_{\vec{K}_\infty}$, i.e. $\sigma\left(\left(f_{\vec{K}_\infty}^x(x)\right); \tau\right) = 0$, which is different from the theta divisor used in the solution of the geodesic equation in Schwarzschild-de Sitter or Reissner-Nordström-de Sitter space-time in the sense that it corresponds to a different normalized period matrix τ. The constant $\varphi_{\text{in}} = \varphi_0 + \int_{u_0}^{\infty} dz_1$ depends only on the initial values $u_0 = \frac{M}{r_0}$ and φ_0.

The case $d = 7$ can be handled analogously to Reissner-Nordström-de Sitter space-time, see (3.3.67). If we apply a substitution $\bar{r} = \frac{1}{\xi} + \bar{r}_P$ where $\bar{r}_P$ is a zero of R_7, we obtain the differential equation

$$\left(\xi\frac{d\xi}{d\varphi}\right)^2 = b_5\xi^5 + \ldots b_0\xi^0 \tag{3.4.8}$$

with some constants b_i. The solution is

$$r(\varphi) = -M\frac{\sigma_2(\vec{\varphi}_{\Theta,7})}{\sigma_1(\vec{\varphi}_{\Theta,7})} + \bar{r}_P = -M\frac{\sigma_2}{\sigma_1}\left(\begin{pmatrix}f_{\vec{K}_\infty}(\varphi - \varphi_{\text{in}}) \\ \varphi - \varphi_{\text{in}}\end{pmatrix}\right) + \bar{r}_P, \tag{3.4.9}$$

where $\sigma(\vec{\varphi}_{\Theta,7}; \tau) = 0$.

3.4. Higher-dimensional space-times

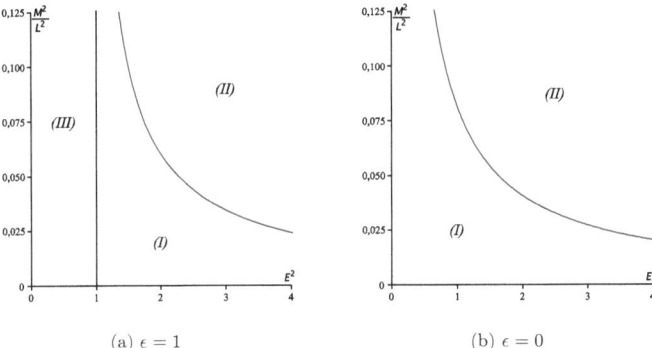

Figure 3.20: Regions of different types of geodesic motion in 6-dimensional Schwarzschild space-time for test particles ($\epsilon = 1$) and light ($\epsilon = 0$). For effective potentials see Fig. 3.21

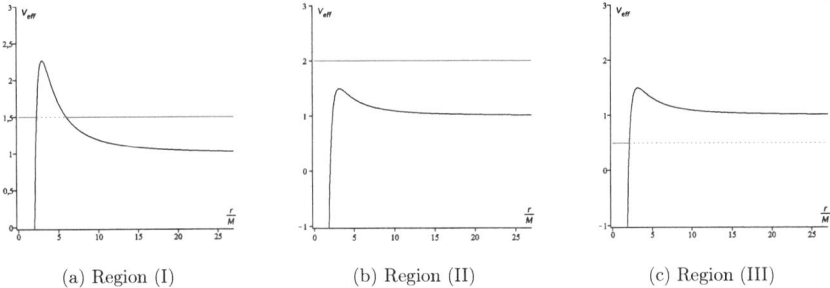

Figure 3.21: Effective potentials for different regions of geodesic motion in 6-dimensional Schwarzschild space-time. The horizontal lines denote the squared energy parameter E^2.

Here, again, $f_{\vec{K}_\infty}$ describes the theta divisor corresponding to the normalized period matrix τ corresponding to the polynomial $\sum_i b_i x^i$, i.e. $\sigma \left(\begin{pmatrix} f_{\vec{K}_\infty}(\varphi - \varphi_{\text{in}}) \\ \varphi - \varphi_{\text{in}} \end{pmatrix} ; \tau \right) = 0$, and $\varphi_{\text{in}} = \varphi_0 + \int_{u_0}^{\infty} dz_2$.

Figure 3.20 shows the arrangement of zeros for a 6-dimensional Schwarzschild space-time. The regions (I)-(III) have the same meaning as in Schwarzschild space-time, i.e.

(I) $R_6(\bar{r})$ has 2 positive real zeros $r_1 < r_2$ with $R_S(\bar{r}) > 0$ for $0 \leq \bar{r} \leq r_1$ and $r_2 \leq \bar{r}$. Possible orbit types: flyby and terminating bound orbits.

(II) $R_6(\bar{r})$ has 0 positive real zeros and $R_6(\bar{r}) \geq 0$ for positive $\bar{r}$. Possible orbit types: terminating escape orbits.

(III) $R_6(\bar{r})$ has 1 positive real zero r_1 with $R_6(\bar{r}) \geq 0$ for $0 \leq \bar{r} \leq r_1$. Possible orbit types:

terminating bound orbits.

There are no bound orbits for any values of E^2 and $\mathcal{L}$ since the polynomial R_6 possesses at most two positive zeros. For light ($\epsilon = 0$) only regions (I) and (II) are possible. For each region corresponding effective potentials are shown in Fig. 3.21.

CHAPTER 4

Geodesics in axially symmetric space-times

In this chapter the results of Chap. 3 are generalized to axially symmetric black hole solutions of the Einstein equations (3.0.1) with and without a cosmological constant. The metric of the gravitational fields considered here have the Boyer-Lindquist form

$$ds^2 = \frac{\Delta_r}{\chi^2 \rho^2}\left(dt - a\sin^2\theta d\varphi\right)^2 - \frac{\Delta_\theta}{\chi^2 \rho^2}\sin^2\theta(adt - (r^2 + a^2)d\varphi)^2 - \frac{\rho^2}{\Delta_\theta}d\theta^2 - \frac{\rho^2}{\Delta_r}dr^2 , \quad (4.0.1)$$

where $\rho^2 = r^2 + a^2 \cos^2\theta$, Δ_r depends only on r, and Δ_θ depends only on θ. The functions Δ_r, Δ_θ, and χ will be specified for the different space-times considered in this chapter, namely the Kerr and Kerr-(anti-)de Sitter space-times as well as briefly, and with some generalizations of (4.0.1), the family of Plebański-Demiański space-times without acceleration of the gravitating object. (Again, units are used where the speed of light c and the gravitational constant G are equal to 1.) The Kerr and Kerr-(anti-)de Sitter space-times are characterized by the mass M, the angular momentum per mass $a = J/M$, and the cosmological constant Λ of the black hole. All space-times with metrics of type (4.0.1) have a singularity at $\rho^2 = 0$, i.e. at simultaneously $r = 0$ and $\theta = \frac{\pi}{2}$. From a transformation of the Boyer-Lindquist form (4.0.1) to cartesian-like coordinates it becomes obvious that $\rho^2 = 0$ corresponds to a ring singularity [14]. Therefore, in these space-times geodesics do not terminate at $r = 0$, $\theta \neq \frac{\pi}{2}$ and negative values of r are possible. However, large negative values of r correspond to a negative mass of the black hole [71]. The case that the ring singularity is naked is widely neglected in this chapter. Other singularities of (4.0.1) located at the axis $\theta = 0, \pi$ and at $\Delta_r = 0$ can be shown to be coordinate singularities [14].

4. Geodesics in axially symmetric space-times

The geodesic motion in such a space-time is described by the geodesic equation

$$0 = \frac{d^2 x^\mu}{ds^2} + \left\{ {}^{\mu}_{\rho\sigma} \right\} \frac{dx^\rho}{ds} \frac{dx^\sigma}{ds} \tag{4.0.2}$$

where $\left\{ {}^{\mu}_{\rho\sigma} \right\}$ is the Christoffel symbol, see (3.0.4).

We start with the derivation of some general properties of geodesics, which are valid for both space-times of type (4.0.1) discussed in the following sections, and proceed in the second section with a review of geodesics in Kerr space-time. The corresponding metric is given by (4.0.1) with [10]

$$\Delta_r = \Delta_{r,K} = r^2 + a^2 - 2Mr,$$
$$\Delta_\theta = 1, \quad \text{and} \quad \chi = 1 \tag{4.0.3}$$

resulting in

$$\begin{aligned}
ds^2 &= \frac{\Delta_{r,K}}{\rho^2} \left(dt - a \sin^2 \theta d\varphi \right)^2 - \frac{\rho^2}{\Delta_{r,K}} dr^2 - \frac{\sin^2 \theta}{\rho^2} (adt - (r^2 + a^2) d\varphi)^2 - \rho^2 d\theta^2 \\
&= \frac{\Delta_{r,K} - a^2 \sin^2 \theta}{\rho^2} dt^2 + \sin^2 \theta \frac{\Delta_{r,K} a^2 \sin^2 \theta - (r^2 + a^2)^2}{\rho^2} d\varphi^2 \\
&\quad + 2a \sin^2 \theta \frac{(r^2 + a^2) - \Delta_{r,K}}{\rho^2} dt d\varphi - \frac{\rho^2}{\Delta_{r,K}} dr^2 - \rho^2 d\theta^2 .
\end{aligned} \tag{4.0.4}$$

Dependent on the parameters M and a of the black hole possible types of orbits in this space-time will be discussed. Following an idea of Mino [30], the geodesic equations in Kerr space-times can be decoupled and an analytical solution can be found in terms of elliptic functions.

In the third section we consider the generalization of the metric discussed in the foregoing section to the case of a nonvanishing cosmological constant. The metric of Kerr-(anti-)de Sitter space-times resulting from (4.0.1) with

$$\Delta_r = \Delta_{r,\text{KdS}} = \left(1 - \frac{\Lambda}{3} r^2 \right) (r^2 + a^2) - 2Mr,$$
$$\Delta_\theta = 1 + \frac{a^2 \Lambda}{3} \cos^2 \theta, \quad \text{and} \quad \chi = 1 + \frac{a^2 \Lambda}{3} \tag{4.0.5}$$

is given by [12]

$$\begin{aligned}
ds^2 &= \frac{\Delta_{r,\text{KdS}}}{\chi^2 \rho^2} \left(dt - a \sin^2 \theta d\varphi \right)^2 - \frac{\rho^2}{\Delta_{r,\text{KdS}}} dr^2 \\
&\quad - \frac{\Delta_\theta \sin^2 \theta}{\chi^2 \rho^2} (adt - (r^2 + a^2) d\varphi)^2 - \frac{\rho^2}{\Delta_\theta} d\theta^2 \\
&= \frac{\Delta_{r,\text{KdS}} - \Delta_\theta a^2 \sin^2 \theta}{\chi^2 \rho^2} dt^2 + \sin^2 \theta \frac{\Delta_{r,\text{KdS}} a^2 \sin^2 \theta - \Delta_\theta (r^2 + a^2)^2}{\chi^2 \rho^2} d\varphi^2 \\
&\quad + 2a \sin^2 \theta \frac{\Delta_\theta (r^2 + a^2) - \Delta_{r,\text{KdS}}}{\chi^2 \rho^2} dt d\varphi - \frac{\rho^2}{\Delta_{r,\text{KdS}}} dr^2 - \frac{\rho^2}{\Delta_\theta} d\theta^2 .
\end{aligned} \tag{4.0.6}$$

As in Kerr space-time, types of orbits will be discussed and, in particular, the influence of the cosmological constant on the geodesic motion will be analyzed. Furthermore, the analytical solutions to the geodesic equations, again decoupled following Mino [30], are derived. This requires the techniques presented in Sec. 3.3 and additionally a method to analytically solve hyperelliptic integrals of third kind. Afterwards, the analytical solution of the geodesic equation will be used to derive analytical expressions for observables.

Finally, it will be demonstrated that the presented methods can even be used to analytically solve the geodesic equation in the general class of Plebański-Demiański space-times without acceleration. As this class of space-times is the most general admitting a separable Hamilton-Jacobi equation, which in turn exhaust all integrable cases [72, 73, 74], it will be shown in this section that in all integrable cases indeed an explicit analytical solution can be given.

4.1 General types of orbits

The geodesic equation (4.0.2) can be separated due to the existence of four constants of motion. Two of these correspond to the energy per unit mass E and the angular momentum per unit mass in z direction L_z given by the generalized momenta p_t and p_φ

$$p_t = g_{tt}\dot{t} + g_{t\varphi}\dot\varphi =: E\,, \tag{4.1.1}$$
$$-p_\varphi = -g_{\varphi\varphi}\dot\varphi - g_{t\varphi}\dot{t} =: L_z\,,$$

where the dot denotes a derivative with respect to the proper time τ. In addition, a third constant of motion is given by the normalization condition $\epsilon = g_{\mu\nu}\dot{x}^\mu \dot{x}^\nu$ with $\epsilon = 1$ for timelike and $\epsilon = 0$ for null geodesics. A fourth constant of motion can be obtained in the process of separation of the Hamilton-Jacobi equation

$$2\frac{\partial S}{\partial \tau} = g^{ij}\frac{\partial S}{\partial x^i}\frac{\partial S}{\partial x^j} \tag{4.1.2}$$

using the ansatz

$$S = \frac{1}{2}\epsilon\tau - Et + L_z\varphi + S_r(r) + S_\theta(\theta)\,. \tag{4.1.3}$$

If we insert this into (4.1.2) we get

$$\epsilon a^2 \cos^2\theta + \Delta_\theta\left(\frac{\partial S_\theta}{\partial \theta}\right)^2 + \frac{\chi^2}{\Delta_\theta \sin^2\theta}\left(aE\sin^2\theta - L_z\right)^2$$
$$= -\epsilon r^2 - \Delta_r\left(\frac{\partial S_r}{\partial r}\right)^2 + \frac{\chi^2}{\Delta_r}\left((r^2+a^2)E - aL_z\right)^2\,, \tag{4.1.4}$$

4. Geodesics in axially symmetric space-times

where each side depends on r or θ only. This means that each side is equal to a constant K, the famous Carter constant, [11]. In the case of $\Delta_\theta = 1$ Eq. (4.1.4) can be reformulated as

$$\epsilon a^2 \cos^2\theta + \left(\frac{\partial S_\theta}{\partial \theta}\right)^2 + \chi^2 \left(\frac{L_z^2}{\sin^2\theta} - a^2 E^2\right)\cos^2\theta$$

$$= -\epsilon r^2 - \Delta_r \left(\frac{\partial S_r}{\partial r}\right)^2 + \frac{\chi^2}{\Delta_r}\left((r^2+a^2)E - aL_z\right)^2 - \chi^2(aE - L_z)^2, \quad (4.1.5)$$

where we used the relation

$$\frac{1}{\sin^2\theta}\left(aE\sin^2\theta - L_z\right)^2 = \left(\frac{L_z^2}{\sin^2\theta} - a^2E^2\right)\cos^2\theta + (aE - L_z)^2. \quad (4.1.6)$$

Both sides in Eq. (4.1.5) are also equal to a constant, the modified Carter constant $\mathcal{C}$, which is related to K by $\mathcal{C} = K - \chi^2(aE - L_z)^2$. In the following, we will use both forms of the Carter constant. Usually, K yields simpler formulas whereas $\mathcal{C}$ can be used for geometrical interpretations (see below).

From the separation ansatz (4.1.3) we derive the equations of motion

$$\rho^4 \left(\frac{dr}{d\tau}\right)^2 = \chi^2((r^2+a^2)E - aL_z)^2 - \Delta_r(\epsilon r^2 + K) =: R(r), \quad (4.1.7)$$

$$\rho^4 \left(\frac{d\theta}{d\tau}\right)^2 = \Delta_\theta(K - \epsilon a^2 \cos^2\theta) - \frac{\chi^2}{\sin^2\theta}(aE\sin^2\theta - L_z)^2 =: \Theta(\theta), \quad (4.1.8)$$

$$\frac{\rho^2}{\chi^2}\frac{d\varphi}{d\tau} = \frac{a}{\Delta_r}((r^2+a^2)E - aL_z) - \frac{1}{\Delta_\theta \sin^2\theta}(aE\sin^2\theta - L_z), \quad (4.1.9)$$

$$\frac{\rho^2}{\chi^2}\frac{dt}{d\tau} = \frac{r^2+a^2}{\Delta_r}((r^2+a^2)E - aL_z) - \frac{a}{\Delta_\theta}(aE\sin^2\theta - L_z). \quad (4.1.10)$$

In the following sections we will explicitly solve these equations for different choices of Δ_r, Δ_θ, and χ. Analogously to the case of spherically symmetric space-times Eq. (4.1.7) suggests the introduction of an effective potential $V_{\text{eff},r}$ such that $V_{\text{eff},r} = E$ corresponds to $\left(\frac{dr}{d\tau}\right)^2 = 0$. However, in contrast to the spherically symmetric case, there are two solutions

$$V_{\text{eff},r}^\pm = \frac{\chi L_z a \pm \sqrt{\Delta_r(\epsilon r^2 + K)}}{(r^2 + a^2)\chi}, \quad (4.1.11)$$

where $\left(\frac{dr}{d\tau}\right)^2 \geq 0$ for $E \leq V_{\text{eff},r}^-$ and $E \geq V_{\text{eff},r}^+$. In the same way an effective potential corresponding to Eq. (4.1.8) can be introduced

$$V_{\text{eff},\theta}^\pm = \frac{L\chi \pm \sqrt{\Delta_\theta \sin^2(\theta)(K - \epsilon a^2 \cos^2(\theta))}}{a\sin^2(\theta)\chi}, \quad (4.1.12)$$

but here $\left(\frac{d\theta}{d\tau}\right)^2 \geq 0$ for $V_{\text{eff},\theta}^- \leq E \leq V_{\text{eff},\theta}^+$.

The following different types of orbits can be identified in space-times described by the metric (4.0.1), see 3.1 and [75].

4.1. General types of orbits

(i) Flyby orbit: $\bar{r}$ starts from $\pm\infty$, then approaches a periapsis $\bar{r} = r_p$ and goes back to $\pm\infty$.

(ii) Bound orbit: $\bar{r}$ oscillates between to boundary values $r_1 \leq \bar{r} \leq r_2$ with $-\infty < r_1 < r_2 < \infty$.

(iii) Transit orbit: $\bar{r}$ starts from $\pm\infty$ and goes to $\mp\infty$ crossing $\bar{r} = 0$.

All other types of orbits are exceptional and treated separately. They are either connected with the ring singularity $\rho^2 = 0$ or with the appearance of multiple zeros on the right hand side of (4.1.7), which simplifies the structure of this differential equation considerably. Examples for the latter type are circular orbits and homoclinic orbits, see [76]. Orbits terminating at the ring singularity can be classified analogously to spherically symmetric space-times:

(iv) Terminating bound orbit: r starts in $(0, r_a]$ with $0 < r_a < \infty$ or in $[r_a, 0)$ with $-\infty < r_a < 0$ and falls into the singularity at $\rho^2 = 0$.

(v) Terminating escape orbit: r comes from $\pm\infty$ and falls into the singularity.

As large negative $\bar{r}$ correspond to a negative mass of the black hole [71], we will assign the attribute *crossover* to flyby, bound, or terminating orbits which pass from positive to negative $\bar{r}$ or vice versa. (By definition, a transit orbit is always a crossover orbit and, therefore, we will not explicitly state that.)

These types of geodesic motion correspond to different arrangements of real zeros of R and Θ together with the sign of R and Θ between them. Thus, for given parameters of the space-time (a, Λ) and the test particle or light ray (ϵ, E, L_z, K) a geodesic motion is possible only in intervals of r and θ where $R(r) \geq 0$ and $\Theta(\theta) \geq 0$. These two conditions can be studied separately.

Let us consider in more detail the meaning of the Carter constant K or, more precisely, of the modified Carter constant $\mathcal{C}$. It will become obvious with the following theorems that $\mathcal{C}$ has a direct geometrical interpretation. The corresponding theorems for the special case of $\Lambda = 0$ can be found in [75].

Theorem 4.1. *If a geodesic lies entirely in the equatorial plane $\theta = \frac{\pi}{2}$ or if it hits the ring singularity $\rho^2 = 0$ then the modified Carter constant $\mathcal{C} = K - \chi^2(aE - L_z)^2$ is zero.*

Proof. A geodesic lies entirely in the equatorial plane iff $\theta(\tau) = \frac{\pi}{2}$ for all τ. This implies that $\Theta(\theta) = (\frac{d\theta}{d\tau})^2 = 0$ and with

$$\Theta\left(\theta = \frac{\pi}{2}\right) = K - \chi^2(aE - L_z)^2 = \mathcal{C}$$

4. Geodesics in axially symmetric space-times

it follows $\mathcal{C} = 0$. If a geodesic hits the ring singularity, then there is a τ such that $r(\tau) = 0$ and $\theta(\tau) = \frac{\pi}{2}$. As $R(r) \geq 0$ and $\Theta(\theta) \geq 0$ for all r and θ of the geodesic, and in particular for $\bar{r} = 0$, $\theta = \frac{\pi}{2}$, it follows

$$R(r=0) = \chi^2 a^2 (aE - L_z)^2 - a^2 K = -a^2 \mathcal{C} \geq 0 \Rightarrow \mathcal{C} \leq 0$$

and as above $\Theta\left(\theta = \frac{\pi}{2}\right) = \mathcal{C} \geq 0$. □

This theorem implies that $\mathcal{C} = 0$ is a necessary condition for equatorial orbits, which is an important class found in many astrophysical objects like accretion discs and planetary systems. Note that the modified Carter constant $\mathcal{C}$ depends on the cosmological constant through $\chi^2 = 1 + \frac{a^2}{3}\Lambda$, which also influences the next theorem.

Theorem 4.2. *For $\Lambda > -\frac{3}{a^2}$ all timelike and null geodesics have $K \geq 0$. In this case $K = 0$ implies $\mathcal{C} = 0$ and the geodesic lies entirely in the equatorial plane.*

Proof. A geodesic can only exist if there are values for $\bar{r}(\tau)$ and $\theta(\tau)$ with $R(r) \geq 0$ and $\Theta(\theta) \geq 0$. From $\Lambda > \frac{-3}{a^2}$ it follows $\Delta_\theta = 1 + \frac{a^2}{3}\Lambda \cos^2\theta > 1 - \cos^2\theta \geq 0$. If $K < 0$ then $(K - \epsilon a^2 \cos^2\theta) < 0$ and

$$\Theta(\theta) = \Delta_\theta(K - \epsilon a^2 \cos^2\theta) - \frac{\chi^2}{\sin^2\theta}(aE\sin^2\theta - L_z)^2 < 0$$

for all values of θ. Assume now $K = 0$. Consequently

$$\Theta(\theta) = -\epsilon a^2 \cos^2\theta \Delta_\theta - \frac{\chi^2}{\sin^2\theta}(aE\sin^2\theta - L_z)^2 \leq 0$$

and $\Theta(\theta) = 0$ only if $\cos^2\theta = 0$ and additionally $(aE\sin^2\theta - L_z) = aE - L_z = 0$. □

Since from observation the cosmological constant has a small positive value, the condition $\Lambda > \frac{-3}{a^2}$ is fulfilled.

From these two theorems it is obvious that, while K originates from the separation procedure, the modified Carter constant $\mathcal{C}$ has a geometric interpretation since it is related to possible θ values of the orbits. This relation can be formulated even more explicite with the help of (i) Θ is symmetric with respect to $\theta = \frac{\pi}{2}$, (ii) Θ has at most 3 real zeros in $[0, \frac{\pi}{2}]$, (iii) $\lim_{\theta \to 0} \Theta(\theta) = -\infty$ (assuming $L_z \neq 0$), and (iv) $\Theta(\frac{\pi}{2})$ has the same sign as $\mathcal{C}$:

- For $\mathcal{C} < 0$ the θ coordinate is confined to $0 < \theta_{\min} \leq \theta \leq \theta_{\max} < \frac{\pi}{2}$ (or the corresponding region in the southern hemisphere).

- If $\Lambda \geq 0$ a positive modified Carter constant $\mathcal{C} > 0$ implies that the θ coordinate oscillates around the equatorial plane. This follows from the fact that in this case: (i) for $\Lambda = 0$ there may be at most 2 real zeros in $[0, \frac{\pi}{2}]$, and (ii) for $\Lambda > 0$ there may be only 1 real zero in $[0, \frac{\pi}{2}]$ as can be inferred from $-\epsilon a^4 \frac{\Lambda}{3} x_1 x_2 x_3 = \mathcal{C}$, where x_i are the zeros of $\Theta(x = \cos(\theta)^2)$.

- The case $\mathcal{C} = 0$ implies that $\theta = \frac{\pi}{2}$ is a double zero of Θ and, thus, that the corresponding geodesic lies entirely in the equatorial plane (stable or unstable), or asymptotically approaches the equatorial plane.

The sign of the modified Carter constant $\mathcal{C}$ also determines whether crossover orbits are possible. The value $r = 0$ is allowed for a geodesic iff

$$0 \leq R(0) = \chi^2 a^2 (aE - L_z)^2 - a^2 K = -a^2 \mathcal{C} (= -a^2 \Theta(\tfrac{\pi}{2})). \tag{4.1.13}$$

It follows that $r = 0$ can only be crossed if $\mathcal{C} < 0$ whereas for $\mathcal{C} > 0$ a transition from positive to negative r is not possible. For $\mathcal{C} = 0$ the geodesic hits the ring singularity if it lies entirely in the equatorial plane or the geodesic asymptotically approaches the singularity.

4.2 Kerr space-time

This section deals with the geodesic equation (4.0.2)

$$0 = \frac{d^2 x^\mu}{ds^2} + \{^{\mu}_{\rho\sigma}\} \frac{dx^\rho}{ds} \frac{dx^\sigma}{ds}$$

in Kerr space-times given by the metric (4.0.4) and characterized by the mass M as well as the angular momentum per mass $a = J/M$ of the black hole. The geodesic motion in Kerr space-time is described by Eqs. (4.1.7) - (4.1.10) with the coordinate functions (4.0.3)

$$\Delta_r = \Delta_{r,K} = r^2 + a^2 - 2Mr,$$
$$\Delta_\theta = 1, \quad \text{and} \quad \chi = 1.$$

The equations of motion (4.1.7) - (4.1.10) are coupled by $\rho^2 = r^2 + \cos^2\theta$. This difficulty can be overcome by introducing the Mino time λ [30] which is related to the proper time τ by $\frac{d\tau}{d\lambda} = \rho^2$. Then

4. Geodesics in axially symmetric space-times

the equations of motions decouple and read

$$\left(\frac{dr}{d\lambda}\right)^2 = R_{\mathrm{K}}(r) = ((r^2 + a^2)E - aL_z)^2 - \Delta_{r,\mathrm{K}}(\epsilon r^2 + K) , \tag{4.2.1}$$

$$\left(\frac{d\theta}{d\lambda}\right)^2 = \Theta_{\mathrm{K}}(\theta) = K - \epsilon a^2 \cos^2\theta - \frac{1}{\sin^2\theta}(aE\sin^2\theta - L_z)^2 , \tag{4.2.2}$$

$$\frac{d\varphi}{d\lambda} = \frac{a}{\Delta_{r,\mathrm{K}}}((r^2 + a^2)E - aL_z) - \frac{1}{\sin^2\theta}(aE\sin^2\theta - L_z) , \tag{4.2.3}$$

$$\frac{dt}{d\lambda} = \frac{r^2 + a^2}{\Delta_{r,\mathrm{K}}}((r^2 + a^2)E - aL_z) - a(aE\sin^2\theta - L_z) . \tag{4.2.4}$$

The function $R_{\mathrm{K}}(r)$ is a polynomial of degree 4 and, thus, Eq. (4.2.1) can in general be solved in terms of elliptic functions. Eq. (4.2.2) can be transformed such that its right hand side is a polynomial of degree 3 and, thus, can also be solved in terms of elliptic functions. These solutions can then be substituted in the remaining two equations (4.2.3) and (4.2.4) yielding elliptic integrals of third kind.

Kerr space-times can be classified according to their number of horizons which depends on the rotation parameter a. Horizons are located at the coordinate singularities given by $\Delta_{r,\mathrm{K}} = r^2 + a^2 - 2Mr = 0$, i.e. at $r_\pm = M \pm \sqrt{M^2 - a^2}$. We speak of *slow* Kerr if there are two horizons, i.e. if $a^2 < M^2$, of *extreme* Kerr if there is one horizon, i.e. if $a^2 = M^2$, and of *fast* Kerr if there is no horizon at all, i.e. if $a^2 > M^2$. For the slow case with $a^2 < M^2$ the outer horizon is an event horizon and the inner a Cauchy horizon. In the following section we will concentrate on the slow case, but the extreme and fast case can be studied analogously. However, the analytical solution to the geodesic equation presented afterwards is valid for all types of Kerr space-times.

4.2.1 Types of orbits

All possible orbit types in Kerr space-time can be read of the right hand sides $R_{\mathrm{K}}(r)$ and $\Theta_{\mathrm{K}}(\theta)$ of Eqs. (4.2.1) and (4.2.2). For given parameters M and a of the black hole, a geodesic with parameters ϵ, E, L_z, and K can exist only if there is any $\theta \in [0, \pi]$ such that $\Theta_{\mathrm{K}}(\theta) \geq 0$ and any r such that $R_{\mathrm{K}}(r) \geq 0$. The orbit type of a geodesic can then be determined by the number of positive real zeros of R_{K} together with the sign of its leading coefficient as explained in Sec. 4.1.

For the analysis of the dependence of the possible types of orbits on the parameters of the space-time and the test particle or light ray it is convenient to rescale the parameters appearing in Eqs. (4.2.1) to (4.2.4) such that they are dimensionless. Thus, we introduce

$$\bar{r} = \frac{r}{M}, \quad \bar{t} = \frac{t}{M}, \quad \bar{a} = \frac{a}{M}, \quad \bar{L}_z = \frac{L_z}{M}, \quad \bar{K} = \frac{K}{M^2}, \tag{4.2.5}$$

and accordingly

$$\Delta_{\bar{r},\mathrm{K}} = \bar{r}^2 + \bar{a}^2 - 2\bar{r}, \quad (\Delta_{r,\mathrm{K}} = M^2 \Delta_{\bar{r},\mathrm{K}}). \tag{4.2.6}$$

In addition, we can absorb M in the definition of λ by introducing

$$\gamma = M\lambda. \tag{4.2.7}$$

Then the equations (4.2.1)-(4.2.4) can be rewritten as

$$\left(\frac{d\bar{r}}{d\gamma}\right)^2 = \mathbb{P}^2(r) - \Delta_{\bar{r},K}(\epsilon \bar{r}^2 + \bar{K}) =: \bar{R}_K(\bar{r}), \tag{4.2.8}$$

$$\left(\frac{d\theta}{d\gamma}\right)^2 = \bar{K} - \epsilon \bar{a}^2 \cos^2\theta - \frac{\mathbb{T}^2(\theta)}{\sin^2\theta} =: \bar{\Theta}_K(\theta), \tag{4.2.9}$$

$$\frac{d\varphi}{d\gamma} = \frac{\bar{a}}{\Delta_{\bar{r},K}}\mathbb{P}(r) - \frac{1}{\sin^2\theta}\mathbb{T}(\theta), \tag{4.2.10}$$

$$\frac{d\bar{t}}{d\gamma} = \frac{\bar{r}^2 + \bar{a}^2}{\Delta_{\bar{r},K}}\mathbb{P}(r) - \bar{a}\mathbb{T}(\theta). \tag{4.2.11}$$

where

$$\begin{aligned}\mathbb{P}(r) &= (\bar{r}^2 + \bar{a}^2)E - \bar{a}\bar{L}_z, \\ \mathbb{T}(\theta) &= \bar{a}E\sin^2\theta - \bar{L}_z.\end{aligned} \tag{4.2.12}$$

First, we will study the motion in θ direction.

Types of latitudinal motion

Geodesics can take an angle θ if and only if $\bar{\Theta}_K(\theta) \geq 0$. Thus, it has to be determined which values of $\bar{a}$, E, $\bar{L}_z$, $\bar{K}$, and $\theta \in [0, \pi]$ result in positive $\bar{\Theta}_K(\theta)$. For simplicity, we substitute $\nu = \cos^2\theta$ giving

$$\bar{\Theta}_K(\nu) = \bar{K} - \epsilon \bar{a}^2 \nu - \left(\bar{a}^2 E^2 (1-\nu) - 2\bar{a}E\bar{L}_z + \frac{\bar{L}_z^2}{1-\nu}\right). \tag{4.2.13}$$

Assume now that for a given set of parameters there exists a certain number of zeros of $\bar{\Theta}_K(\nu)$ in $[0, 1]$. If we vary the parameters, the position of zeros varies and the number of real zeros in $[0, 1]$ can change only if (i) a zero crosses 0 or 1 or (ii) two zeros merge. Let us consider case (i). 0 is a zero iff

$$\bar{\Theta}_K(\nu = 0) = \bar{K} - (\bar{a}E - \bar{L}_z)^2 = 0 \tag{4.2.14}$$

or

$$\bar{L}_z = \bar{a}E \pm \sqrt{\bar{K}}. \tag{4.2.15}$$

As $\nu = 1$ is in general a pole of $\bar{\Theta}_K(\nu)$ it is a necessary condition for 1 being a zero of $\bar{\Theta}_K(\nu)$ that this pole becomes a removable singularity. From (4.2.13) it follows that this is the case for $\bar{L}_z = 0$. Under this assumption we obtain

$$\bar{\Theta}_K(\nu) = \bar{a}^2(E^2 - \epsilon)\nu + \bar{K} - \bar{a}^2 E^2 \quad \text{for } \bar{L}_z = 0 \tag{4.2.16}$$

$$\stackrel{\nu=1}{=} \bar{K} - \epsilon \bar{a}^2. \tag{4.2.17}$$

4. Geodesics in axially symmetric space-times

Summarized, $\bar{L}_z = \bar{a}E \pm \sqrt{\bar{K}}$ and simultaneously $\bar{L}_z = 0$ and $\bar{K} = \epsilon\bar{a}^2$ give us border cases of the θ motion.

Now let us consider case (ii). If we exclude the coordinate singularities $\theta = 0, \pi$ or $\nu = 1$ the zeros of $\bar{\Theta}_K(\nu)$ are given by the zeros of

$$\Theta_\nu(\nu) = (1-\nu)(\bar{K} - \epsilon\bar{a}^2\nu) - (\bar{a}E - \bar{L}_z - \bar{a}E\nu)^2, \qquad (4.2.18)$$
$$= -\bar{a}^2(E^2 - \epsilon)\nu^2 + (2\bar{a}E(\bar{a}E - \bar{L}_z) - \bar{K} - \epsilon\bar{a}^2)\nu + \bar{K} - (\bar{a}E - \bar{L}_z)^2,$$

which is in general a polynomial of degree 2. The two zeros of Θ coincide at a double zero $x \in [0,1)$ iff

$$\Theta_\nu(x) = 0 \quad \text{and} \quad \frac{d\Theta_\nu}{d\nu}(x) = 0. \qquad (4.2.19)$$

These two equations can be solved for $\bar{L}_z$ depending on E and the remaining parameters $\bar{a}$ and $\bar{K}$ by

$$\bar{L}_z = \frac{(E \pm \sqrt{E^2 - 1})(\bar{a}^2 - \bar{K})}{2\bar{a}} \quad \text{for} \quad \epsilon = 1,$$
$$\bar{L}_z = \frac{-\bar{K}}{4\bar{a}E} \quad \text{for} \quad \epsilon = 0. \qquad (4.2.20)$$

From this equation it is obvious that in the case of timelike geodesics double zeros may only occur for $E^2 \geq 1$. More precisely, from the condition that double zeros are considered only in $[0,1)$ a lower bound for E can be derived, see below. The representation (4.2.20) of values of $\bar{L}_z$ again correspond to border cases of the θ motion. Let us additionally consider the conditions for $\nu = 1$ being a double zero. With $\bar{L}_z = 0$ and $\bar{K} = \epsilon\bar{a}^2$ it follows

$$\left.\frac{d\Theta_K(\nu)}{d\nu}\right|_{\nu=1} = \bar{a}^2(E^2 - \epsilon), \qquad (4.2.21)$$

which is zero for $E^2 = \epsilon$. Note that in the case $\bar{L}_z = 0$ the function $\Theta_K(\nu)$ is linear and, thus, that $\bar{L}_z = 0$, $\bar{K} = \epsilon\bar{a}^2$, and $E^2 = \epsilon$ correspond to $\Theta_K \equiv 0$.

From the number of zeros of Θ_K in $[0,1]$ the type of latitudinal motion can be inferred if the sign of Θ_K at $\theta = 0$ (which is same as at $\theta = \pi$) is known. For $\bar{L}_z \neq 0$, it is $\lim_{\theta \to 0} \Theta_K(\theta) = -\infty$, but for $\bar{L}_z = 0$ the sign of $\Theta_K(\theta = 0)$ is given by $\bar{K} - \epsilon\bar{a}^2$.

For given parameter of the black hole $\bar{a}$ we can use these informations to analyse the θ motion of all possible geodesics in this space-time. As a typical example for timelike geodesics in a slow Kerr space-time consider Fig. 4.1, where the curves divide the half plane into four regions (a)-(d) which correspond to different arrangement of zeros in $[0,1]$ and to the following different types of motion in θ direction:

(a) no geodesic motion possible,

(b) Θ_ν has one real zero $\nu_{\max}$ in $[0,1)$ with $\Theta_\nu \geq 0$ for $\nu \in [0, \nu_{\max}]$, i.e. θ oscillates around the equatorial plane $\theta = \frac{\pi}{2}$,

(c) no geodesic motion possible,

(d) Θ_ν has two real zeros $\nu_{\min}, \nu_{\max}$ in $[0,1)$ with $\Theta_\nu \geq 0$ for $\nu \in [\nu_{\min}, \nu_{\max}]$, i.e. θ oscillates between $\arccos(\pm\sqrt{\nu_{\min}})$ and $\arccos(\pm\sqrt{\nu_{\max}})$.

A geodesic motion is only possible in regions (b) and (d) because in all other regions Θ_ν is negative for all $\nu \in [0,1]$. Note that for the special case of $\bar{K} = \epsilon \bar{a}^2$ strictly speaking regions (b) and (d) are divided by $\bar{L}_z = 0$ as this is a border case of the θ motion. However, in each region we have for $\bar{L}_z > 0$ and $\bar{L}_z < 0$ the same number of zeros in $[0,1]$ and, thus, the same type of motion. (More precisely, near $\bar{L}_z = 0$ a zero $\nu_0 < 1$ of $\Theta_K(\nu)$ approaches 1, but does not cross it.) Therefore, in each region we put the parts above and below $\bar{L}_z = 0$ together.

Note that the case $\bar{L}_z = 0$ itself has to be treated separately as Θ_ν is not the appropriate function here but $\Theta_K(\nu)$, which is then a linear function with a zero at $\frac{\bar{a}^2 E^2 - \bar{K}}{\bar{a}^2(E^2 - \epsilon)}$. If $\frac{\bar{K}}{\bar{a}^2} < \epsilon$, the case $\bar{L}_z = 0$ belongs to region (b) if $E^2 \leq \frac{\bar{K}}{\bar{a}^2}$ but to region (a) or (c) if $E^2 > \frac{\bar{K}}{\bar{a}^2}$. However, if $\frac{\bar{K}}{\bar{a}^2} > \epsilon$, θ may take all values in $[0,\pi]$ if $E^2 < \frac{\bar{K}}{\bar{a}^2}$ but only a range $0 \leq \theta \leq \theta_{\max}$ (or $\pi \geq \theta \geq \pi - \theta_{\max}$) if $E^2 \geq \frac{\bar{K}}{\bar{a}^2}$. In the case of $\frac{\bar{K}}{\bar{a}^2} = \epsilon$, $\theta = 0, \pi$ is always a zero of Θ_K and $\Theta_K(\frac{\pi}{2}) = -\bar{a}^2(E^2 - \epsilon)$.

Typical examples of the corresponding effective potentials for each region are pictured in Fig. 4.2. Remember that $\Theta_K > 0$ between the two potentials $V^\pm_{\text{eff},\theta}$, see (4.1.12). The boundaries of region (b) are given by $\bar{L}_z = \bar{a}E \pm \sqrt{\bar{K}}$ (for both $\epsilon = 0$ and $\epsilon = 1$) and, therefore, the regions gets larger if $\bar{K}$ grows. A change of $\bar{a}$ in addition causes region (b) to shift up or down. The upper boundary of (d) is identical with the lower boundary of region (b) whereas the lower boundary of (d) is given by (4.2.20). The point where the upper and lower boundaries of region (d) touch each other is where θ is a double zero of Θ_ν, which is given by

$$E = \frac{1}{2}\frac{\bar{K} + \bar{a}^2}{\sqrt{\bar{K}}\bar{a}} \quad \text{for} \quad \epsilon = 1,$$

$$E = \frac{1}{2}\frac{\sqrt{\bar{K}}}{\bar{a}} \quad \text{for} \quad \epsilon = 0. \quad (4.2.22)$$

The regions (b) and (d) are characterized in a simple way in terms of the modified Carter constant $\mathcal{C}$ as explained in Sec. 4.1: As in region (d) $\Theta_\nu(0) < 0$ it follows that this region corresponds to $\mathcal{C} < 0$. In the same way we can conclude that region (b), where $\Theta_\nu(0) > 0$, corresponds to $\mathcal{C} > 0$. In addition, crossover orbits corresponding to $\mathcal{C} < 0$ are only possible in region (d).

4. Geodesics in axially symmetric space-times

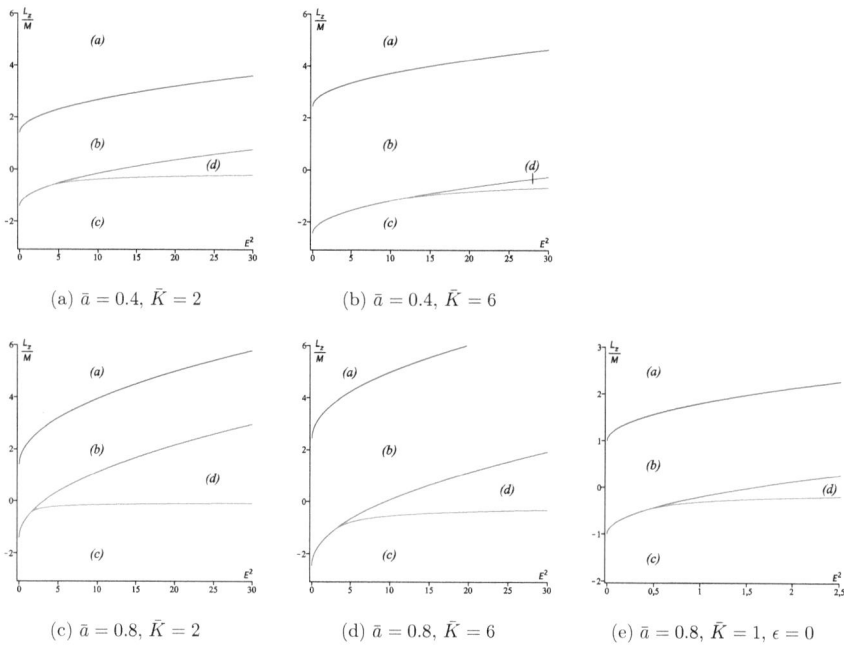

Figure 4.1: Typical regions of different types of θ-motion in slow Kerr space-time. A geodesic motion is only possible in the regions (b) and (d) in each plot. On the boundaries of region (b) the modified Carter constant $\mathcal{C}$ vanishes. Here $\epsilon = 1$ with the exception of plot (e), where it is shown that region (d) may have $E^2 < 1$ for $\epsilon = 0$ (note the rescaled axes). For effective potentials see Fig. 4.2.

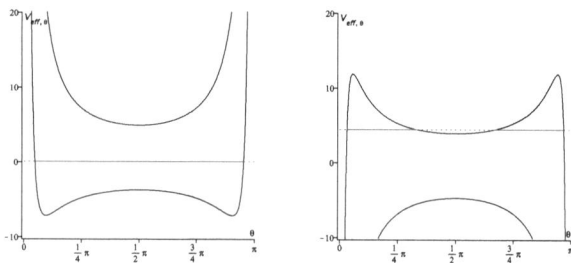

Figure 4.2: Effective potentials for different regions of the theta motion in Kerr space-time. The horizontal lines denote the energy parameter E.

Types of radial motion

A geodesic can take a radial coordinate $\bar{r}$ if and only if

$$\bar{R}_{\mathrm{K}}(\bar{r}) = ((\bar{r}^2 + \bar{a}^2)E - \bar{a}\bar{L}_z)^2 - \Delta_{\bar{r},\mathrm{K}}(\epsilon \bar{r}^2 + \bar{K}) \geq 0.$$

The zeros of $\bar{R}_{\mathrm{K}}$ are extremal values of $\bar{r}(\gamma)$ and determine the type of geodesic. The polynomial $\bar{R}_{\mathrm{K}}$ is in general of degree 4 in $\bar{r}$ and, therefore, has 4 possibly complex zeros of which the real zeros are of interest for the type of motion. As a Kerr space-time has no singularity in $\bar{r} = 0, \theta \neq \frac{\pi}{2}$, we can also consider negative $\bar{r}$ as valid.

For a given set of parameters, $\bar{R}_{\mathrm{K}}$ has a certain number of real zeros. If we vary the parameters this number can change only if two zeros merge to one. This happens at $\bar{r} = x$ iff

$$R_{\mathrm{K}}(\bar{r}) = (\bar{r} - x)^2(a_2\bar{r}^2 + a_1\bar{r} + a_0) \tag{4.2.23}$$

for some real constants a_i. By a comparison of coefficients we can solve the resulting 5 equations for $E^2(x)$ and $\bar{L}_z(x)$ depending on the remaining parameters $\bar{a}$ and $\bar{K}$. This yields two complicated expressions for E^2 and $\bar{L}_z$, which can not be analyzed easily for the influence of each of the parameters $\bar{a}$ and $\bar{K}$. However, a typical result in a slow Kerr space-time including the results of the analysis of the θ motion is shown in Fig. 4.3 for timelike geodesics and in 4.4 for null geodesics. Here, the boundary of region (I) correspond to pairs of E^2 and $\bar{L}_z$ which are given by negative x, the vertical line to $E^2 = \epsilon$ (where $\lim_{\bar{r}\to\infty} \bar{R}_{\mathrm{K}}(\bar{r})$ changes from plus to minus infinity), the remaining solid line to positive x, and the dashed lines to the analysis of the latitudinal motion. The alterations due to changing $\bar{a}$ and $\bar{K}$ are shown in Fig. 4.5. From these plots it can be concluded that for growing $\bar{a}$ and $\bar{K}$ regions (IIIb) and (IVb) are enlarged. As in general for growing $\bar{K}$ region (b) of the θ motion is widened, it can be inferred that in this case the regions (IIb) to (Vb) are enlarged for fixed $\bar{a}$. However, at the same time the regions (Id) and (IId) gets smaller.

We recognize five regions of different types of timelike r motion. (Here we always assume $r_i < r_{i+1}$.)

(I) All zeros of R_{K} are complex and $\bar{R}_{\mathrm{K}}(\bar{r}) \geq 0$ for all $\bar{r}$. Possible orbit types: transit orbit.

(II) $\bar{R}_{\mathrm{K}}$ has two real zeros r_1, r_2 and $\bar{R}_{\mathrm{K}}(\bar{r}) \geq 0$ for $\bar{r} \leq r_1$ and $r_2 \leq \bar{r}$. Possible orbit types: two flyby orbits, one to $+\infty$ and one to $-\infty$.

(III) All four zeros r_i, $1 \leq i \leq 4$, of $\bar{R}_{\mathrm{K}}$ are real and $\bar{R}_{\mathrm{K}}(\bar{r}) \geq 0$ for $r_{2k-1} \leq \bar{r} \leq r_{2k}$, $k = 1, 2$. Possible orbit types: two different bound orbits.

(IV) Again all four zeros of $\bar{R}_{\mathrm{K}}$ are real but $\bar{R}_{\mathrm{K}}(\bar{r}) \geq 0$ for $\bar{r} \leq r_1$, $r_2 \leq \bar{r} \leq r_3$, and $r_4 \leq \bar{r}$. Possible orbit types: two flyby orbits, one to each of $\pm\infty$ and a bound orbit.

4. Geodesics in axially symmetric space-times

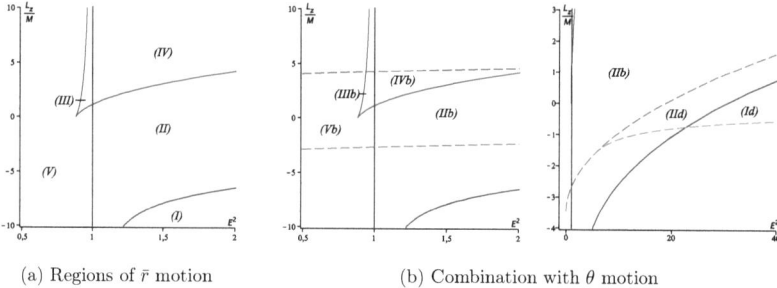

(a) Regions of $\bar{r}$ motion

(b) Combination with θ motion

Figure 4.3: Regions of different types of timelike geodesics in Kerr space-time for both $\bar{r}$ and θ motion. Here $\bar{a} = 0.8$ and $\bar{K} = 12$. Solid lines correspond to boundaries of the $\bar{r}$ motion and dashed lines to boundaries of the θ motion.

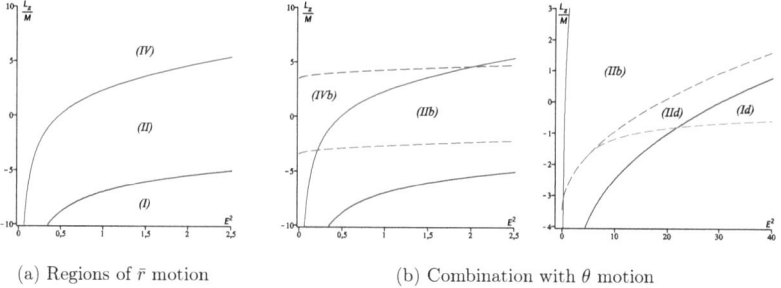

(a) Regions of $\bar{r}$ motion

(b) Combination with θ motion

Figure 4.4: Regions of different types of null geodesics in Kerr space-time for both $\bar{r}$ and θ motion. Here $\bar{a} = 0.8$ and $\bar{K} = 12$. Solid lines correspond to boundaries of the $\bar{r}$ motion and dashed lines to boundaries of the θ motion.

(V) $\bar{R}_K$ has two real zeros r_1, r_2 and $\bar{R}_K(\bar{r}) \geq 0$ for $r_1 \leq \bar{r} \leq r_2$. Possible orbit types: a bound orbit.

Although there is the same number of real zeros the different orbit types in regions (III)/(IV) and (II)/(V) are due to the different behavior of $\bar{R}_K$ when $\bar{r} \to \pm\infty$. The expression $\bar{R}_K = \sum_{i=1}^{4} a_i \bar{r}^i$ is a polynomial of degree 4 with $a_4 = E^2 - \epsilon$ which for $\bar{r} \to \pm\infty$ yields $\bar{R}_K(\bar{r}) \to \infty$ if $E^2 > \epsilon$ and $\bar{R}_K(\bar{r}) \to -\infty$ if $E^2 < \epsilon$. For light ($\epsilon = 0$) only regions (I),(II), and (IV) are present, see Fig. 4.4. The corresponding effective potentials for every region are pictured in Fig. 4.6

Let us also analyse where we have crossover orbits. As region (I) can only contain a transit orbit, which is by definition a crossover orbit, it can only intersect region (d). Examples of $\bar{a}$ and $\bar{K}$ where region (IIb) as well as region (IId) appears are shown in Fig. 4.3. The regions (III) and (V) have

4.2. Kerr space-time

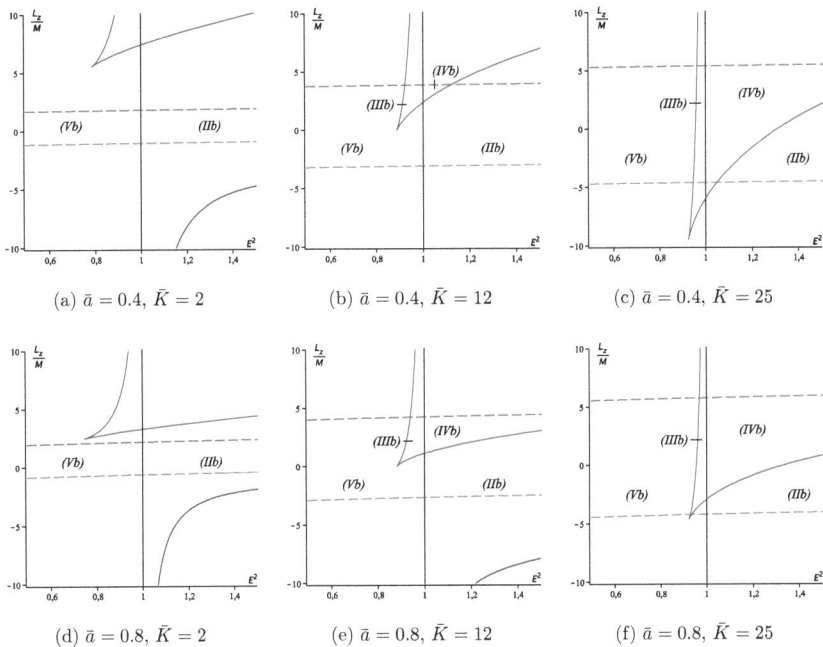

Figure 4.5: Alterations of regions of different types of timelike geodesics in Kerr space-time for varying a and K. Solid lines correspond to boundaries of the $\bar{r}$ motion and dashed lines to boundaries of the θ motion.

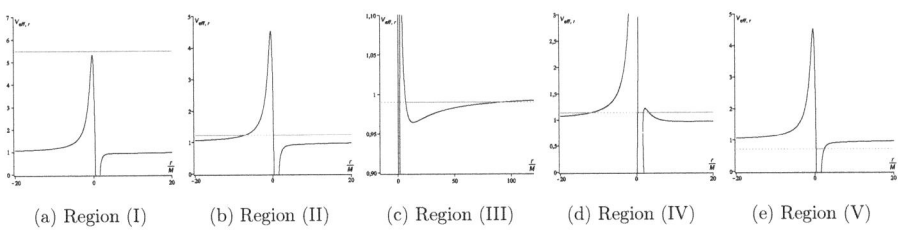

Figure 4.6: Effective potentials for different regions of the r motion in Kerr space-time. The horizontal lines denote the energy parameter E.

81

4. Geodesics in axially symmetric space-times

region	−	+	range of $\bar{r}$	types of orbits
Id	0	0	——————→	transit
IIb	1	1	•——•——→	2x flyby
IId	2	0	•——•——→	flyby, crossover flyby
IIIb	0	4	——•••••——→	2x bound
IVb	1	3	•——•••••——→	2x flyby, bound
Vb	0	2	——•••——→	bound

Table 4.1: Orbit types in slow Kerr space-time. The + and − columns give the number of positive and negative real zeros of the polynomial $\bar{R}_{\rm K}$. In the fourth column, the thick lines represent the range of orbits. Turning points are shown by thick dots. The small vertical line denotes $\bar{r}=0$.

$E^2 < 1$ and, thus, no points in region (d) due to (4.2.22). Therefore, there are no crossover orbits in these regions. Apparently, also region (IV) only contains region (b) of θ motion although this remains to be shown. The results of this paragraph together with the numbers of positive and negative zeros for each region are summarized in Tab. 4.1.

In Figs. 4.7 and 4.8 typical examples for timelike and null geodesics in a slow Kerr space-time are shown. Due to the singularities on the right hand sides of the differential equations (4.2.10) and (4.2.11) for φ and $\bar{t}$ a geodesic approaching one of the horizons at $1 \pm \sqrt{1-\bar{a}^2}$ will infinitely many times spiral around the black hole as well as it will take an infinite time to cross the horizon. However, an observer traveling along such a geodesic will experience a cross of the horizon in a finite time. For describing this smooth cross other coordinates in place of φ and t has to be introduced which have no singularity at that horizon. These coordinates are known as *K or K* coordinates, see [75, 14]. As in Reissner-Nordström space-time, there exist bound and flyby orbits crossing the Cauchy horizon at $\bar{r}_{\rm C} = 1 - \sqrt{1-\bar{a}^2}$, then being reflected and again crossing $\bar{r}_{\rm C}$ thereby entering a new copy of the Kerr space-time. This can be inferred from the Carter-Penrose diagram of the Kerr space-time, see e.g. [14]. We will call such orbits many-world bound or two-world flyby orbits, as in Reissner-Nordström space-time.

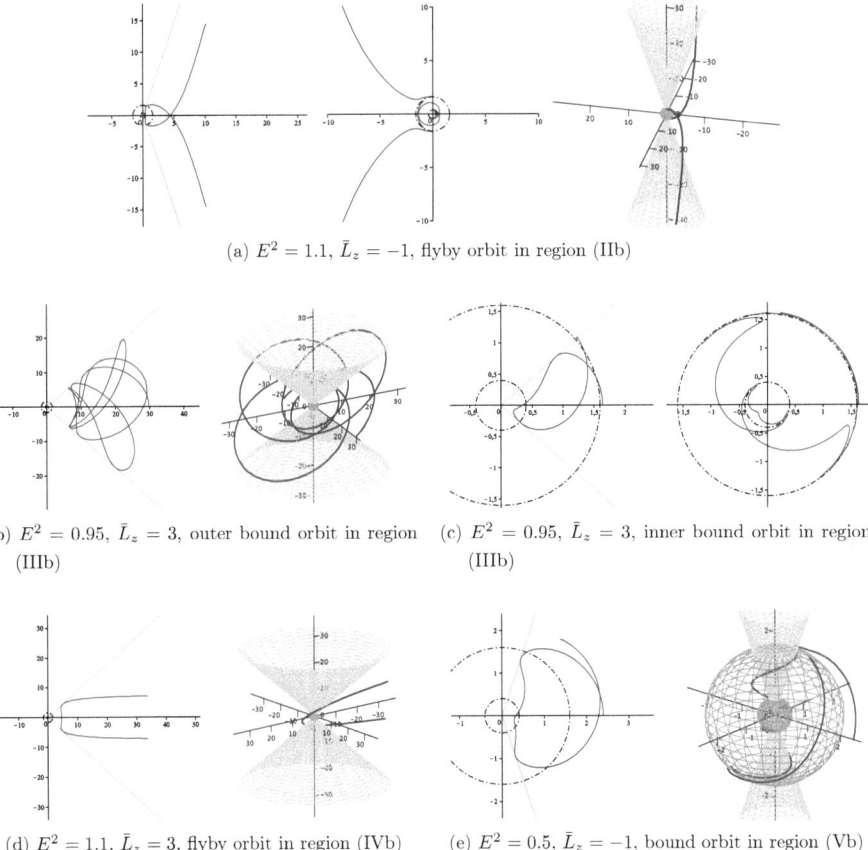

Figure 4.7: Timelike geodesics in Kerr space-time for $\bar{a} = 0.8$ and $\bar{K} = 12$. In each subplot, on the left side the r-θ plane and on the right side a 3d-image and/or the r-φ plane is shown. Light grey lines and cones correspond to extremal θ and dark grey spheres as well as dashed black circles to horizons. The bound orbits in (c) and (e) crosses the Cauchy horizon several times and, thus, are many world bound orbits. The flyby orbit in (a) is a two-world orbit. At every horizon, the φ-coordinate goes to ∞ as witnessed by a distant observer.

4. Geodesics in axially symmetric space-times

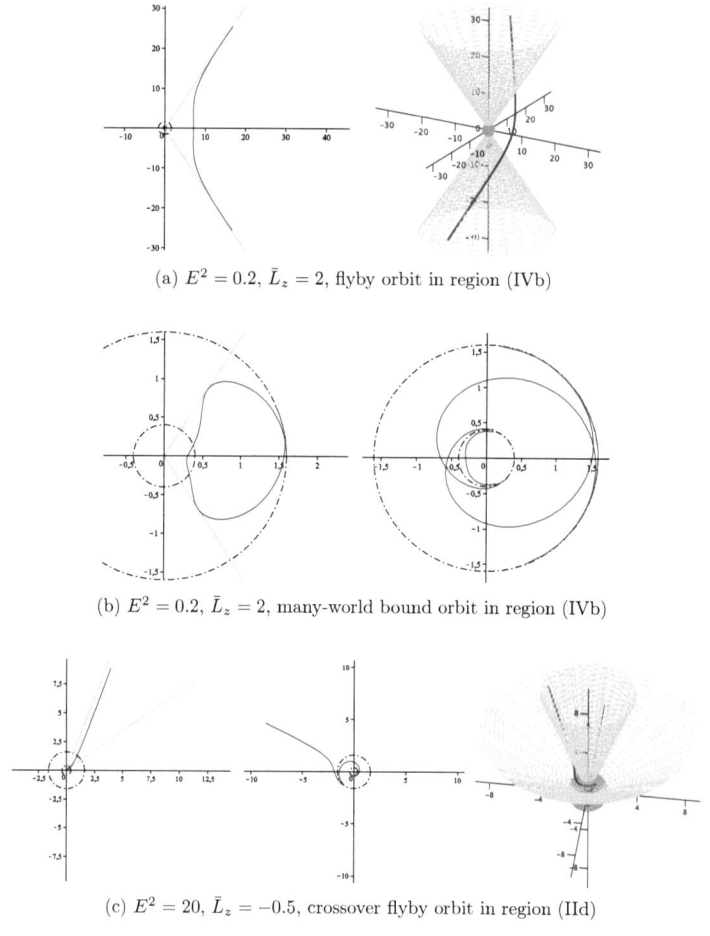

(a) $E^2 = 0.2$, $\bar{L}_z = 2$, flyby orbit in region (IVb)

(b) $E^2 = 0.2$, $\bar{L}_z = 2$, many-world bound orbit in region (IVb)

(c) $E^2 = 20$, $\bar{L}_z = -0.5$, crossover flyby orbit in region (IId)

Figure 4.8: Null geodesics in Kerr space-time for $\bar{a} = 0.8$ and $\bar{K} = 12$. In each subplot, on the left side the r-θ plane and on the right side a 3d-image and/or the r-φ plane is shown. Light grey or black lines, cones, and spheres as in Fig. 4.7.

4.2.2 Analytical solutions of the geodesic equations

In this subsection the analytical solutions of the geodesic equations (4.2.8)-(4.2.11)

$$\left(\frac{d\bar{r}}{d\gamma}\right)^2 = \bar{R}_K(\bar{r}) = \mathbb{P}^2(r) - \Delta_{\bar{r},K}(\epsilon\bar{r}^2 + \bar{K}), \tag{4.2.8}$$

$$\left(\frac{d\theta}{d\gamma}\right)^2 = \bar{\Theta}_K(\theta) = \bar{K} - \epsilon\bar{a}^2\cos^2\theta - \frac{\mathbb{T}^2(\theta)}{\sin^2\theta}, \tag{4.2.9}$$

$$\frac{d\varphi}{d\gamma} = \frac{\bar{a}}{\Delta_{\bar{r},K}}\mathbb{P}(r) - \frac{1}{\sin^2\theta}\mathbb{T}(\theta), \tag{4.2.10}$$

$$\frac{d\bar{t}}{d\gamma} = \frac{\bar{r}^2 + \bar{a}^2}{\Delta_{\bar{r},K}}\mathbb{P}(r) - \bar{a}\mathbb{T}(\theta), \tag{4.2.11}$$

where

$$\mathbb{P}(r) = (\bar{r}^2 + \bar{a}^2)E - \bar{a}\bar{L}_z,$$
$$\mathbb{T}(\theta) = \bar{a}E\sin^2\theta - \bar{L}_z,$$

are presented. Each equation is treated separately.

$\bar{r}$ motion

As mentioned before the function $\bar{R}_K(\bar{r})$ in (4.2.8)

$$\bar{R}_K(\bar{r}) = \mathbb{P}^2(r) - \Delta_{\bar{r},K}(\epsilon\bar{r}^2 + \bar{K})$$

is a polynomial of degree 4 for both $\epsilon = 1$ and $\epsilon = 0$ and, therefore, the differential equation (4.2.8) is of elliptic type if $\bar{R}_K$ has only simple zeros but can be solved in terms of elementary functions if $\bar{R}_K$ has multiple zeros. In the latter case the analytical solution can be found for example in [31]. In the following it is assumed that $\bar{R}_K$ has only simple zeros.

Analogous to the situation in Reissner-Nordström space-time, see Sec. 3.2.2, the polynomial $\bar{R}_K$ can be transformed to the Weierstrass form (2.1.5) by the standard substitutions, i.e. first $\bar{r} = \frac{1}{\xi} + \bar{r}_K$ for a zero $\bar{r}_K$ of $\bar{R}_K(\bar{r})$, what yields

$$\left(\frac{d\xi}{d\gamma}\right)^2 = \sum_{j=0}^{3} a_j \xi^j, \quad a_j = \frac{1}{(4-j)!}\frac{d^{(4-j)}\bar{R}_K}{d\bar{r}^{4-j}}(r_K). \tag{4.2.24}$$

Second, $\xi = \frac{1}{a_3}\left(4y - \frac{a_2}{3}\right)$ casts (4.2.24) in the form

$$\left(\frac{dy}{d\gamma}\right)^2 = 4y^3 - g_2 y - g_3, \tag{4.2.25}$$

4. Geodesics in axially symmetric space-times

where g_2, g_3 are given by the standard equation (2.1.11)

$$g_2 = \frac{1}{16}\left(\frac{4}{3}a_2^2 - 4a_1 a_3\right),$$

$$g_3 = \frac{1}{16}\left(\frac{1}{3}a_1 a_2 a_3 - \frac{2}{27}a_2^3 - a_0 a_3^2\right).$$

The analytical solution of Eq. (4.2.8) for Kerr space-time is then given by

$$\bar{r}(\gamma) = \frac{1}{\xi(\gamma)} + r_K = \frac{a_3}{4\wp(\gamma - \gamma_{r,\text{in}}) - \frac{a_2}{3}} + r_K, \qquad (4.2.26)$$

where

$$\gamma_{r,\text{in}} = \gamma_0 + \int_{y_0}^{\infty} \frac{dz}{\sqrt{4z^3 - g_2 z - g_3}}, \quad y_0 = \frac{1}{4}\left(\frac{a_3}{\bar{r}_0 - r_K} + \frac{a_2}{3}\right), \qquad (4.2.27)$$

depends only on the initial values γ_0 and $\bar{r}_0$.

θ motion

The first step to an analytical solution of Eq. (4.2.9) is a substitution $\nu = \cos^2\theta$ which yields the easier differential equation

$$\left(\frac{d\nu}{d\gamma}\right)^2 = 4\nu\Theta_\nu(\nu) \qquad (4.2.28)$$

$$= -4\bar{a}^2(E^2 - \epsilon)\nu^3 + 4(2\bar{a}E(\bar{a}E - \bar{L}_z) - \bar{K} - \epsilon\bar{a}^2)\nu^2 + 4(\bar{K} - (\bar{a}E - \bar{L}_z)^2)\nu$$

where Θ_ν is defined in (4.2.18). The polynomial on the right hand side of (4.2.28) is in general of degree 3 for both $\epsilon = 1$ and $\epsilon = 0$ and, thus, of elliptic type. However, if the degree of $4\nu\Theta_\nu$ is reduced to 2, which is the case for $E^2 = \epsilon$, or if $\nu\Theta_\nu$ has multiple zeros Eq. (4.2.28) can be solved in terms of elementary functions, see e.g. [31]. In the following we assume that this is not the case.

Eq. (4.2.28) can be cast in the Weierstrass form (2.1.5) with the standard substitution

$$\nu = \frac{3y - 2\bar{a}E(\bar{a}E - \bar{L}_z) + \bar{K} + \epsilon\bar{a}^2}{-3\bar{a}^2(E^2 - \epsilon)}. \qquad (4.2.29)$$

The solution of (4.2.9) is then given by

$$\theta(\gamma) = \arccos\left(\pm\sqrt{\frac{3\wp(\gamma - \gamma_{\theta,\text{in}}) - 2\bar{a}E(\bar{a}E - \bar{L}_z) + \bar{K} + \epsilon\bar{a}^2}{-3\bar{a}^2(E^2 - \epsilon)}}\right), \qquad (4.2.30)$$

where

$$\gamma_{\theta,\text{in}} = \gamma_0 + \int_{y_0}^{\infty} \frac{dz}{\sqrt{4z^3 - g_2 z - g_3}},$$

$$y_0 = \frac{1}{3}\left(-3\bar{a}^2\cos^2\theta_0(E^2 - \epsilon) + 2\bar{a}E(\bar{a}E - \bar{L}_z) - \bar{K} - \epsilon\bar{a}^2\right), \qquad (4.2.31)$$

and
$$g_2 = \frac{4}{3}\left(2\bar{a}E(\bar{a}E - \bar{L}_z) - \bar{K} - \epsilon\bar{a}^2\right)^2 + 4\bar{a}^2(\bar{K} - (\bar{a}E - \bar{L}_z)^2)(E^2 - \epsilon),$$
$$g_3 = 4(2\bar{a}E(\bar{a}E - \bar{L}_z) - \bar{K} - \epsilon\bar{a}^2)\left[\frac{-\bar{a}^2}{3}(\bar{K} - (\bar{a}E - \bar{L}_z)^2)(E^2 - \epsilon)\right. \quad (4.2.32)$$
$$\left. - \frac{2}{27}(2\bar{a}E(\bar{a}E - \bar{L}_z) - \bar{K} - \epsilon\bar{a}^2)^2\right].$$

The sign of the square root in (4.2.30) has to be chosen according to θ being in the northern (plus sign) or southern (minus sign) hemisphere and, thus, depends on the initial value θ_0. If the motion oscillates around the equatorial plane, the two solutions with different signs have to be glued together along $\theta = \frac{\pi}{2}$ starting with the sign indicated by θ_0.

φ motion

The equation for φ
$$\frac{d\varphi}{d\gamma} = \frac{\bar{a}}{\Delta_{\bar{r},K}}\mathrm{P}(r) - \frac{1}{\sin^2\theta}\mathrm{T}(\theta)$$
can be splitted in a part only dependent on r and a part only dependent on θ. An integration yields
$$\varphi - \varphi_0 = \int_{\gamma_0}^{\gamma}\frac{\bar{a}\mathrm{P}(r)}{\Delta_{\bar{r}(\gamma),K}}d\gamma - \int_{\gamma_0}^{\gamma}\frac{\mathrm{T}(\theta)d\gamma}{\sin^2\theta(\gamma)}$$
$$= \int_{\bar{r}_0}^{\bar{r}}\frac{\bar{a}\mathrm{P}(r)d\bar{r}}{\Delta_{\bar{r},K}\sqrt{R_K(\bar{r})}} - \int_{\theta_0}^{\theta}\frac{\mathrm{T}(\theta)d\theta}{\sin^2\theta\sqrt{\Theta_K(\theta)}}, \quad (4.2.33)$$
where we substituted $\bar{r} = \bar{r}(\gamma)$, i.e. $\frac{d\bar{r}}{d\gamma} = \sqrt{R_K}$, in the first and $\theta = \theta(\gamma)$, i.e. $\frac{d\theta}{d\gamma} = \sqrt{\Theta_K}$, in the second integral.

We will solve now the two integrals in (4.2.33) separately.

$\bar{r}$ dependent part Let us consider the first, $\bar{r}$ dependent integral in (4.2.33)
$$I_r := \int_{\bar{r}_0}^{\bar{r}}\frac{\bar{a}\left((\bar{r}^2 + \bar{a}^2)E - \bar{a}\bar{L}_z\right)d\bar{r}}{\Delta_{\bar{r},K}\sqrt{R_K}}. \quad (4.2.34)$$

Analogous to subsection 4.2.2 this integral can be solved by elementary functions if $\bar{R}_K$ has multiple zeros [31]. However, in general I_r is an elliptic integral of third kind. Thus, a transformation of $\bar{R}_K$ to the standard Weierstrass form has to be carried out, what can be achieved in the same way as in subsection 4.2.2. The substitutions $\bar{r} = \xi^{-1} + \bar{r}_K$ for a zero $\bar{r}_K$ and $\xi = \frac{1}{a_3}\left(4y - \frac{a_2}{3}\right)$ where a_3, a_2 are

4. Geodesics in axially symmetric space-times

given by (4.2.24) together with a subsequent partial fraction decomposition cast I_r in the form

$$I_r = -\bar{a}\frac{|a_3|}{a_3}\left[\frac{\mathbb{P}(\bar{r}_K)}{\Delta_{\bar{r}_K,K}}\int_{y_0}^{y}\frac{dy}{\sqrt{P_W(y)}} + \sum_{i=1}^{2}\int_{y_0}^{y}\frac{C_i dy}{(y-y_i)\sqrt{P_W(y)}}\right]$$

$$= -\bar{a}\frac{|a_3|}{a_3}\left[\frac{\mathbb{P}(\bar{r}_K)}{\Delta_{\bar{r}_K,K}}(v-v_0) + \sum_{i=1}^{2}\int_{v_0}^{v}\frac{C_i dv}{(\wp(v)-y_i)}\right], \quad (4.2.35)$$

where $P_W(y) = 4y^3 - g_2 y - g_3$, y_i are the zeros of the polynomial $144\Delta_{\bar{r}_K,K}z^2 - 24(a_2\Delta_{\bar{r}_K,K} + 3a_3(\bar{r}_K - 1))z + a_2^2\Delta_{\bar{r}_K,K} - 6a_2a_3(\bar{r}_K - 1) + 9a_3^2$, and C_i is the coefficient of the partial fraction $(y - y_i)^{-1}$. This integral can now be solved with the standard procedure described in Thm. 2.5 and appendix A. The result is

$$I_r(\gamma) = -\bar{a}\frac{|a_3|}{a_3}\left[\frac{\mathbb{P}(\bar{r}_K)}{\Delta_{\bar{r}_K,K}}(\gamma - \gamma_0)\right.$$
$$\left. + \sum_{i,j=1}^{2}\frac{C_i}{\wp'(v_{ij})}\left(\zeta(v_{ij})(\gamma - \gamma_0) + \log\sigma(\gamma - \gamma_{ij}) - \log\sigma(\gamma_0 - \gamma_{ij})\right)\right], \quad (4.2.36)$$

where $\wp(\gamma_{ij} - \gamma_{r,in}) = y_i$.

θ dependent part We solve now the θ dependent integral

$$I_\theta := \int_{\theta_0}^{\theta}\frac{\left(\bar{a}E\sin^2\theta - \bar{L}_z\right)d\theta}{\sin^2\theta\sqrt{\Theta}}, \quad (4.2.37)$$

which can be transformed to the easier form

$$I_\theta = \mp\int_{\nu_0}^{\nu}\frac{\bar{a}E(1-\nu) - \bar{L}_z}{(1-\nu)\sqrt{4\nu\Theta_\nu}}d\nu \quad (4.2.38)$$

by the substitution $\nu = \cos^2\theta$, where Θ_ν is defined in (4.2.18). Here we have to pay special attention to the integration path. If $\theta \in (0, \frac{\pi}{2}]$ we have $\cos\theta = +\sqrt{\nu}$ but for $\theta \in [\frac{\pi}{2}, \pi)$ it is $\cos\theta = -\sqrt{\nu}$. Accordingly, we first have to split the integration path from θ_0 to θ such that every piece is fully contained in the interval $(0, \frac{\pi}{2}]$ or $[\frac{\pi}{2}, \pi)$ and then to choose the appropriate sign of the square root of ν in $\cos\theta = \sqrt{\nu}$. By these considerations it seems like the sign of I_θ and, thus, the φ coordinate depends on whether the particle is in the nothern or southern hemisphere. But in fact the branches of the square root $\sqrt{4\nu\Theta_\nu}$ also change when the substitution $\cos\theta = +\sqrt{\nu}$ changes to $\cos\theta = -\sqrt{\nu}$. Therefore, in the whole the sign of I_θ does not depend on whether θ is in the northern or southern hemisphere. In the following we assume for simplicity that $\cos\theta = +\sqrt{\nu}$.

Analogous to subsection 4.2.2 the integral I_θ can be solved by elementary functions if $\nu\Theta_\nu$ has at least a double zero [31]. If $\nu\Theta_\nu$ has only simple zeros I_θ is of elliptic type and of third kind. If this is

4.2. Kerr space-time

the case, a substitution

$$\nu = \frac{1}{a_3}\left(4y - \frac{a_2}{3}\right) \qquad (4.2.39)$$

with $\quad a_2 = 4(2\bar{a}E(\bar{a}E - \bar{L}_z) - \bar{K} - \epsilon\bar{a}^2), \quad a_3 = -4\bar{a}^2(E^2 - \epsilon)$

transforms I_θ to

$$\begin{aligned}
I_\theta &= -\frac{|a_3|}{a_3}\int_{y_0}^{y}\frac{4\bar{a}Ey - \bar{a}E(a_3 + \frac{a_2}{3}) + \bar{L}_z a_3}{(4y - a_3 - \frac{a_2}{3})\sqrt{P_W(y)}}dy \\
&= -\frac{|a_3|}{a_3}\left[\bar{a}E\int_{y_0}^{y}\frac{dy}{\sqrt{P_W(y)}} + \int_{y_0}^{y}\frac{a_3\bar{L}_z dy}{(4y - a_3 - \frac{a_2}{3})\sqrt{P_W(y)}}\right] \\
&= -\frac{|a_3|}{a_3}\left[\bar{a}E(v - v_0) + \int_{v_0}^{v}\frac{a_3\bar{L}_z dv}{4\wp(v) - a_3 - \frac{a_2}{3}}\right],
\end{aligned} \qquad (4.2.40)$$

where $P_W(y) = 4y^3 - g_2 y - g_3$ with g_2, g_3 as in (4.2.31) and $\wp(v) = y$. The square root of P_W has to be chosen such that it coincides with the sign of $\wp'$. The integral in (4.2.40) can now be solved with the standard procedure described in Thm. 2.5 and appendix A. The result is

$$\begin{aligned}
I_\theta(\gamma) = -\frac{|a_3|}{a_3}\Bigg[&\bar{a}E(\gamma - \gamma_0) \\
&+ \frac{a_3\bar{L}_z}{4}\sum_{j=1}^{2}\frac{1}{\wp'(v_j)}\bigg(\zeta(v_j)(\gamma - \gamma_0) + \log\sigma(\gamma - \gamma_j) - \log\sigma(\gamma_0 - \gamma_j)\bigg)\Bigg],
\end{aligned} \qquad (4.2.41)$$

where $\wp(\gamma_j - \gamma_{\theta,in}) = \frac{a_3}{4} + \frac{a_2}{12}$.

t motion

The differential equation for $\bar{t}$,

$$\frac{d\bar{t}}{d\gamma} = \frac{\bar{r}^2 + \bar{a}^2}{\Delta_{\bar{r},K}}\mathbb{P}(\bar{r}) - \bar{a}\mathbb{T}(\theta),$$

can be solved analogously to the equation for the φ motion (4.2.10). First, the equation can be integrated yielding

$$\begin{aligned}
\bar{t} - \bar{t}_0 &= \int_{\gamma_0}^{\gamma}\frac{\bar{r}^2 + \bar{a}^2}{\Delta_{\bar{r},K}}\mathbb{P}(\bar{r})d\gamma - \bar{a}\int_{\gamma_0}^{\gamma}\mathbb{T}(\theta)d\gamma \\
&= \int_{\bar{r}_0}^{\bar{r}}\frac{\bar{r}^2 + \bar{a}^2}{\Delta_{\bar{r},K}}\frac{\mathbb{P}(\bar{r})}{\sqrt{R_K(\bar{r})}}d\bar{r} - \bar{a}\int_{\theta_0}^{\theta}\frac{\mathbb{T}(\theta)}{\sqrt{\Theta_K(\theta)}}d\theta \qquad (4.2.42)\\
&=: \tilde{I}_r - \bar{a}\tilde{I}_\theta. \qquad (4.2.43)
\end{aligned}$$

As the solution procedure for the type of integrals on the right hand side of this equation was already explained, we just write down here the result for the most general cases.

4. Geodesics in axially symmetric space-times

The integral $\tilde{I}_r$ is solved by

$$\tilde{I}_r = C_0 \int_{y_0}^{y} \frac{dy}{\sqrt{4y^3 - g_2 y - g_3}} + \sum_{i=1}^{3} C_{i1} \int_{y_0}^{y} \frac{dy}{(y - y_i)\sqrt{4y^3 - g_2 y - g_3}}$$

$$+ C_{32} \int_{y_0}^{y} \frac{dy}{(y - y_3)^2 \sqrt{4y^3 - g_2 y - g_3}}$$

$$= C_0(\gamma - \gamma_0) + \sum_{i=1}^{3} \sum_{j=1}^{2} \frac{C_{i1}}{\wp'(v_{ij})} \left(\zeta(v_{ij})(\gamma - \gamma_0) + \log \sigma(\gamma - \gamma_{ij}) - \log \sigma(\gamma_0 - \gamma_{ij})\right)$$

$$- \sum_{j=1}^{2} \frac{C_{32}}{\wp'(v_{3j})} \left[\frac{\wp(v_{3j})\wp'(v_{3j}) + 2\wp''(v_{3j})\zeta(v_{3j})}{\wp'(v_{3j})^2}(\gamma - \gamma_0) + \zeta(\gamma - \gamma_{3j}) \right.$$

$$\left. + 2\wp''(v_{3j})(\log \sigma(\gamma - \gamma_{3j}) - \log \sigma(\gamma_0 - \gamma_{3j})) \right], \qquad (4.2.44)$$

where C_0 is a constant, C_{ik} are the coefficients of the partial fractions $(y - y_i)^{-k}$ which may be computed with the help of a computer algebra system, y_1, y_2 are the two zeros of $\Delta_{y(\bar{r}),\mathrm{K}}$, $y_3 = \frac{a_2}{12}$ with a_2 as in (4.2.24), g_2, g_3 as in the Sec. 4.2.2, $\wp(v_{i1}) = y_i = \wp(v_{i2})$, and $\gamma_{ij} = \gamma_{r,\mathrm{in}} + v_{ij}$.

With the same substitutions as in Sec. 4.2.2 the polynomial $\Theta(\theta)$ can be transformed to the standard Weierstrass form. With these substitutions the integral $\tilde{I}_\theta$ becomes

$$\tilde{I}_\theta = -\frac{4\bar{a}E}{a_3} \int_{y_0}^{y} \frac{y \, dy}{\sqrt{4y^3 - g_2 y - g_3}} + \frac{3\bar{a}Ea_3 + \bar{a}Ea_2 - 3La_3}{3a_3} \int_{y_0}^{y} \frac{dy}{\sqrt{4y^3 - g_2 y - g_3}}$$

$$= \frac{4\bar{a}E}{a_3}\zeta(\gamma - \gamma_0) + \frac{3\bar{a}Ea_3 + \bar{a}Ea_2 - 3La_3}{3a_3}(\gamma - \gamma_0) \qquad (4.2.45)$$

where $a_3 = -4\bar{a}^2(E^2 - \epsilon)$ and $a_2 = 4(2\bar{a}E(\bar{a}E - \bar{L}_z) - \bar{K} - \epsilon\bar{a}^2)$.

4.3 Kerr-de Sitter space-time

The aim of this section is the generalization of the results of Sec. 4.2 to the case of a nonvanishing cosmological constant. Thus, the geodesic equations (4.1.7)-(4.1.10) in Kerr-(anti-)de Sitter space-time

$$\rho^4 \left(\frac{dr}{d\tau}\right)^2 = R(r) = \chi^2((r^2 + a^2)E - aL_z)^2 - \Delta_{r,\mathrm{KdS}}(\epsilon r^2 + K), \qquad (4.1.7)$$

$$\rho^4 \left(\frac{d\theta}{d\tau}\right)^2 = \Theta(\theta) = \Delta_\theta(K - \epsilon a^2 \cos^2\theta) - \frac{\chi^2}{\sin^2\theta}(aE\sin^2\theta - L_z)^2, \qquad (4.1.8)$$

$$\frac{\rho^2}{\chi^2}\frac{d\varphi}{d\tau} = \frac{a}{\Delta_{r,\mathrm{KdS}}}((r^2 + a^2)E - aL_z) - \frac{1}{\Delta_\theta \sin^2\theta}(aE\sin^2\theta - L_z), \qquad (4.1.9)$$

$$\frac{\rho^2}{\chi^2}\frac{dt}{d\tau} = \frac{r^2 + a^2}{\Delta_{r,\mathrm{KdS}}}((r^2 + a^2)E - aL_z) - \frac{a}{\Delta_\theta}(aE\sin^2\theta - L_z), \qquad (4.1.10)$$

4.3. Kerr-de Sitter space-time

where $\Delta_{r,\text{KdS}}$ and Δ_θ are given by (4.0.5)

$$\Delta_r = \Delta_{r,\text{KdS}} = \left(1 - \frac{\Lambda}{3}r^2\right)(r^2 + a^2) - 2Mr,$$

$$\Delta_\theta = 1 + \frac{a^2\Lambda}{3}\cos^2\theta, \qquad \chi = 1 + \frac{a^2\Lambda}{3},$$

are analyzed and analytically solved along the lines of [77, 78]. The Kerr-(anti-) de Sitter space-time in Boyer-Lindquist form describes an axially symmetric and stationary vacuum solution of Einstein's equations and is characterized by the mass M of the gravitating body, the angular momentum per mass $a = J/M$, and the cosmological constant Λ.

The geodesic equations (4.1.7)-(4.1.10) are coupled by $\rho^2 = r^2 + a^2 \cos^2\theta$. Analogously to Kerr space-time this can be handled by introducing the Mino time λ [30] connected to the proper time τ by $\frac{d\tau}{d\lambda} = \rho^2$. Dependent on this Mino time the equations of motions read

$$\left(\frac{dr}{d\lambda}\right)^2 = R_{\text{KdS}}(r) = \chi^2((r^2+a^2)E - aL_z)^2 - \Delta_{r,\text{KdS}}(\epsilon r^2 + K), \qquad (4.3.1)$$

$$\left(\frac{d\theta}{d\lambda}\right)^2 = \Theta_{\text{KdS}}(\theta) = \Delta_\theta(K - \epsilon a^2 \cos^2\theta) - \frac{\chi^2}{\sin^2\theta}(aE\sin^2\theta - L_z)^2, \qquad (4.3.2)$$

$$\frac{1}{\chi^2}\frac{d\varphi}{d\lambda} = \frac{a}{\Delta_{r,\text{KdS}}}((r^2+a^2)E - aL_z) - \frac{1}{\Delta_\theta \sin^2\theta}(aE\sin^2\theta - L_z), \qquad (4.3.3)$$

$$\frac{1}{\chi^2}\frac{dt}{d\lambda} = \frac{r^2+a^2}{\Delta_{r,\text{KdS}}}((r^2+a^2)E - aL_z) - \frac{a}{\Delta_\theta}(aE\sin^2\theta - L_z). \qquad (4.3.4)$$

As R_{KdS} is in general a polynomial of degree 6, the differential equation (4.3.1) is of hyperelliptic type and an analytical solution can be found with the method presented in Chap. 3, Sec. 3.3. The right hand side Θ_{KdS} of Eq. (4.3.2) can be transformed to a polynomial of degree 4 and, thus, this equation is of elliptic type with a analytical solution in terms of Weierstrass functions. However, Eqs. (4.3.3) and (4.3.4) have the most complicated form considered in this book. If the solutions of (4.3.1) and (4.3.2) are substituted in these two equations the resulting integrals are of hyperelliptic type and third kind.

Analogously to the situation in Kerr space-time, we classify this form of the metric according to the number of (disconnected) regions where $\Delta_{r,\text{KdS}} > 0$, which depends on the parameters M, a and Λ. We speak of *slow* Kerr-de Sitter if there are two regions and of *fast* Kerr-de Sitter if there is one region where $\Delta_{r,\text{KdS}} > 0$. The limiting case where two regions are connected by a zero of $\Delta_{r,\text{KdS}}$ is called *extreme* Kerr-de Sitter. Other cases are not possible, what can be seen by a comparison of coefficients in $\Delta_{r,\text{KdS}} = -\frac{\Lambda}{3}r^4 + (1 - \frac{\Lambda}{3}a^2)r^2 - 2Mr + a^2 = -\frac{\Lambda}{3}\prod_{i=1}^{4}(r - r_i)$ where r_i denote the zeros of $\Delta_{r,\text{KdS}}$. Fig. 4.9 shows the modification of regions of slow, fast, and extreme Kerr-de Sitter with varying Λ. From this plot it can be inferred that for a positive cosmological constant the gravitating object in a slow Kerr-de Sitter space-time is not necessarily slowly rotating with respect to its mass.

4. Geodesics in axially symmetric space-times

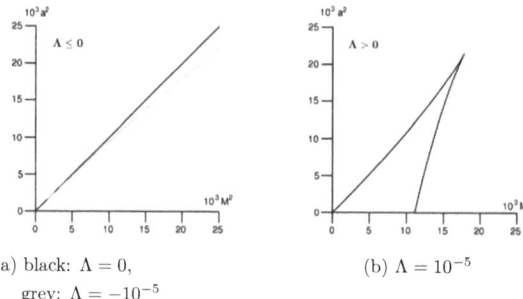

(a) black: $\Lambda = 0$,
grey: $\Lambda = -10^{-5}$

(b) $\Lambda = 10^{-5}$

Figure 4.9: Regions of slow and fast Kerr-de Sitter for different values of Λ. a) Below the line we have a slow Kerr-anti-de Sitter space-time (with a Cauchy and an event horizon) and above a fast one corresponding to a naked singularity. b) In the region bounded by the two curves we have a slow Kerr-de Sitter space-time with two cosmological horizons, where one is located at negative r, a Cauchy horizon and an event horizon. Outside we have a fast one with two cosmological horizons, one again at negative r.

4.3.1 Types of orbits

Before solving the equations of motion for Kerr-de Sitter space-time, we analyse the structure of possible orbits dependent on the black hole parameters $\bar{a}$, $\bar{\Lambda}$ and the particle parameters ϵ, E, L_z, and K. As explained in Sec. 4.1 the major point in this analysis is that (4.3.1) and (4.3.2) imply $R_{\text{KdS}}(\bar{r}) \geq 0$ and $\Theta_{\text{KdS}}(\theta) \geq 0$ as a necessary condition for the existence of a geodesic.

Again, it is most convenient to introduce dimensionless quantities for an analysis of the dependence of the possible types of orbits on the parameters of the space-time and the geodesic. Thus, we define

$$\bar{r} = \frac{r}{M}, \quad \bar{t} = \frac{t}{M}, \quad \bar{a} = \frac{a}{M}, \quad \bar{L}_z = \frac{L_z}{M}, \quad \bar{K} = \frac{K}{M^2}, \quad \bar{\Lambda} = \frac{1}{3}\Lambda M^2 \qquad (4.3.5)$$

and accordingly

$$\Delta_{\bar{r},\text{KdS}} = \left(1 - \bar{\Lambda}\bar{r}^2\right)(\bar{r}^2 + \bar{a}^2) - 2\bar{r}, \quad (\Delta_{r,\text{KdS}} = M^2 \Delta_{\bar{r},\text{KdS}}),$$
$$\Delta_\theta = 1 + \bar{a}^2 \bar{\Lambda} \cos^2 \theta, \quad \text{and} \quad \chi = 1 + \bar{a}^2 \bar{\Lambda}. \qquad (4.3.6)$$

In addition, we can absorb M in the definition of λ by introducing

$$\gamma = M\lambda. \qquad (4.3.7)$$

Then the equations (4.3.1)-(4.3.4) can be rewritten as

$$\left(\frac{d\bar{r}}{d\gamma}\right)^2 = \chi^2 \mathbb{P}^2(r) - \Delta_{\bar{r},\text{KdS}}(\epsilon \bar{r}^2 + \bar{K}) =: \bar{R}_{\text{KdS}}(\bar{r}), \tag{4.3.8}$$

$$\left(\frac{d\theta}{d\gamma}\right)^2 = \Delta_\theta(\bar{K} - \epsilon \bar{a}^2 \cos^2\theta) - \frac{\chi^2 \mathbb{T}^2(\theta)}{\sin^2\theta} =: \bar{\Theta}_{\text{KdS}}(\theta), \tag{4.3.9}$$

$$\frac{1}{\chi^2} \frac{d\varphi}{d\gamma} = \frac{\bar{a}}{\Delta_{\bar{r},\text{KdS}}} \mathbb{P}(r) - \frac{1}{\Delta_\theta \sin^2\theta} \mathbb{T}(\theta), \tag{4.3.10}$$

$$\frac{1}{\chi^2} \frac{d\bar{t}}{d\gamma} = \frac{\bar{r}^2 + \bar{a}^2}{\Delta_{\bar{r},\text{KdS}}} \mathbb{P}(r) - \frac{\bar{a}}{\Delta_\theta} \mathbb{T}(\theta), \tag{4.3.11}$$

where

$$\mathbb{P}(r) = (\bar{r}^2 + \bar{a}^2)E - \bar{a}\bar{L}_z,$$
$$\mathbb{T}(\theta) = \bar{a} E \sin^2\theta - \bar{L}_z,$$

as in (4.2.12). In the remainder of this section we will study the consequences of the two conditions $R_{\text{KdS}}(\bar{r}) \geq 0$ and $\Theta_{\text{KdS}}(\theta) \geq 0$.

Types of latitudinal motion

Analogously to the case of a vanishing cosmological constant the function $\Theta_{\text{KdS}}(\theta)$ can be simplified by introducing $\nu = \cos^2\theta$, which yields

$$\Theta_{\text{KdS}}(\nu) = (1 + \bar{a}^2 \bar{\Lambda} \nu)(\bar{K} - \epsilon \bar{a}^2 \nu) - \chi^2 \left(\bar{a}^2 E^2 (1-\nu) - 2\bar{a} E \bar{L}_z + \frac{\bar{L}_z^2}{1-\nu}\right). \tag{4.3.12}$$

It will be analyzed in the following which values of $\bar{a}, \bar{\Lambda}, E, \bar{L}_z, \bar{K}$, and $\theta \in [0,\pi]$ result in positive $\Theta_{\text{KdS}}(\nu)$. For this, assume that for a given set of parameters there exists a certain number of zeros of $\Theta_{\text{KdS}}(\nu)$ in $[0,1]$. These zeros correspond to turning points of the θ motion. Therefore, the type of θ motion changes if the number of real zeros in $[0,1]$ changes. This may happen only if (i) a zero crosses 0 or 1 or (ii) two zeros merge to one. Case (i) occurs iff

$$\Theta_{\text{KdS}}(\nu = 0) = \bar{K} - \chi^2(\bar{a}E - \bar{L}_z)^2 = 0 \quad \Leftrightarrow \quad \bar{L}_z = \bar{a}E \pm \frac{\sqrt{\bar{K}}}{\chi} \tag{4.3.13}$$

or $\Theta_{\text{KdS}}(\nu = 1) = 0$. As $\nu = 1$ is in general a pole of $\Theta_{\text{KdS}}(\nu)$ it is a necessary condition for $\Theta_{\text{KdS}}(\nu = 1) = 0$ that this pole becomes a removable singularity. From (4.3.12) it can be inferred that this happens for $\bar{L}_z = 0$. If this is inserted in (4.3.12) we obtain

$$\Theta_{\text{KdS}}(\nu) = (1 + \bar{a}^2 \bar{\Lambda} \nu)(\bar{K} - \epsilon \bar{a}^2 \nu) - \chi^2 \bar{a}^2 E^2 (1-\nu) \quad \text{for } \bar{L}_z = 0 \tag{4.3.14}$$
$$\stackrel{\nu=1}{=} \chi(\bar{K} - \epsilon \bar{a}^2).$$

4. Geodesics in axially symmetric space-times

Thus, it can be inferred that $\Theta_{\text{KdS}}(\nu = 1) = 0$ if and only if $\bar{L}_z = 0$ and additionally $\bar{K} = \epsilon \bar{a}^2$ (as $\chi = 0$ can be excluded, see also Thm. 4.2). Summarized, $\bar{L}_z = \bar{a}E \pm \frac{\sqrt{\bar{K}}}{\chi}$ and simultaneously $\bar{L}_z = 0$ and $\bar{K} = \epsilon \bar{a}^2$ specify border cases of the θ motion.

Now let us study case (ii). If the coordinate singularities $\theta = 0, \pi$ or $\nu = 1$ are excluded the zeros of $\Theta_{\text{KdS}}(\nu)$ are given by the zeros of

$$\Theta_\nu = (1 - \nu)(1 + \bar{a}^2 \bar{\Lambda} \nu)(\bar{K} - \epsilon \bar{a}^2 \nu) - \chi^2 \left(\bar{a}E(1 - \nu) - \bar{L}_z\right)^2, \qquad (4.3.15)$$

which is in general a polynomial of degree 3 for $\epsilon = 1$ but of degree 2 for $\epsilon = 0$. Then for timelike geodesics two zeros coincide at $x \in [0, 1)$ iff

$$\Theta_\nu = (\nu - x)^2 (a_1 \nu + a_0) \qquad (4.3.16)$$

for some real constants a_1, a_0. By a comparison of coefficients we can solve the resulting 4 equation for $\bar{L}_z(x)$ and $E^2(x)$ depending on the remaining parameters $\bar{a}, \bar{\Lambda}$, and $\bar{K}$. This parametric representation of values of $\bar{L}_z$ and E^2 again correspond to border cases of the θ motion. For null geodesics a parametric representation can be obtained by setting $a_1 = 0$ in (4.3.16), or by directly comparing the only two zeros of Θ_ν, which coincide iff

$$\bar{L}_z = \frac{\chi E \pm \sqrt{\bar{K}\bar{\Lambda} + \chi^2 E^2}}{2 \bar{a} \bar{\Lambda}}. \qquad (4.3.17)$$

Let us additionally consider the conditions for $\nu = 1$ being a double zero. With $\bar{L}_z = 0$ it follows

$$\left.\frac{d\Theta_{\text{KdS}}(\nu)}{d\nu}\right|_{\nu=1} = \bar{a}^2 \bar{\Lambda}(\bar{K} - \epsilon \bar{a}^2) + \bar{a}^2 \chi (\chi E^2 - \epsilon), \qquad (4.3.18)$$

where additionally $\bar{K} = \epsilon \bar{a}^2$ has to be fulfilled. Then $\nu = 1$ is a double zero for $E^2 = \frac{\epsilon}{\chi}$.

To determine the type of latitudinal motion, in addition to the number of zeros in $[0, \pi]$ the sign of Θ_{KdS} at $\theta = 0$ and $\theta = \pi$ has to be known. In the case of $\bar{L}_z \neq 0$, this sign is given by $\lim_{\theta \to 0} \Theta_{\text{KdS}}(\theta) = -\infty$, but for $\bar{L}_z = 0$ it is $\Theta_{\text{K}}(\theta = 0) = \chi(\bar{K} - \epsilon \bar{a}^2)$.

For given parameters $\bar{a}, \bar{\Lambda}$ of the black hole, we can use these informations to identify all possible types of θ motion in this space-time. As a typical example for timelike geodesics consider Fig. 4.10. Analogously to the situation for a vanishing cosmological constant the curves divide the half plane into four regions (a)-(d) which correspond to different arrangement of zeros in $[0, 1]$. Again, geodesic motion is only possible in regions (b) and (d) because in all other regions $\bar{\Theta}_{\text{KdS}}$ is negative for all $\nu \in [0, 1]$. The types of θ motion in the regions (b) and (d) are the same as in a Kerr space-time, i.e.:

(b) Here Θ_ν has one real zero ν_{max} in $[0, 1]$ with $\Theta_\nu \geq 0$ for $\nu \in [0, \nu_{\text{max}}]$, i.e. θ oscillates around the equatorial plane $\theta = \frac{\pi}{2}$.

4.3. Kerr-de Sitter space-time

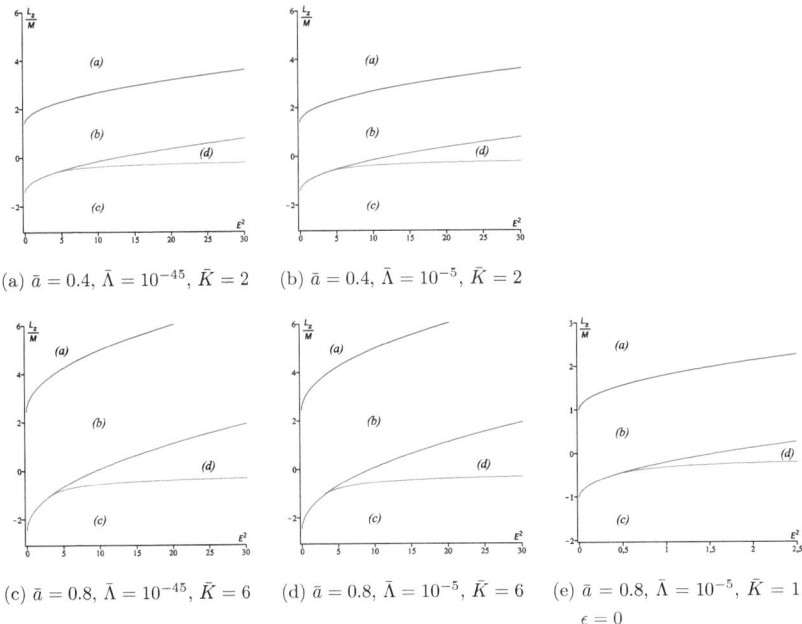

(a) $\bar{a} = 0.4$, $\bar{\Lambda} = 10^{-45}$, $\bar{K} = 2$ (b) $\bar{a} = 0.4$, $\bar{\Lambda} = 10^{-5}$, $\bar{K} = 2$

(c) $\bar{a} = 0.8$, $\bar{\Lambda} = 10^{-45}$, $\bar{K} = 6$ (d) $\bar{a} = 0.8$, $\bar{\Lambda} = 10^{-5}$, $\bar{K} = 6$ (e) $\bar{a} = 0.8$, $\bar{\Lambda} = 10^{-5}$, $\bar{K} = 1$, $\epsilon = 0$

Figure 4.10: Typical regions of different types of θ-motion in slow Kerr-de Sitter space-time. A geodesic motion is only possible in the regions (b) and (d) in each plot. On the boundaries of region (b) the modified Carter constant $\mathcal{C}$ vanishes. Here $\epsilon = 1$ with the exception of plot (e), where it is shown that region (d) may have $E^2 < 1$ for $\epsilon = 0$ (note the rescaled axes). For effective potentials see Fig. 4.2.

(d) In this case Θ_ν has two real zeros $\nu_{\min}, \nu_{\max}$ in $[0,1]$ with $\Theta_\nu \geq 0$ for $\nu \in [\nu_{\min}, \nu_{\max}]$, i.e. θ oscillates between $\arccos(\pm\sqrt{\nu_{\min}})$ and $\arccos(\pm\sqrt{\nu_{\max}})$.

As in Kerr space-time, the case of $\bar{L}_z = 0$ has to be treated separately because Θ_{KdS} instead of Θ_ν has to be used here for the analysis.

Compared to the case of $\bar{\Lambda} = 0$ with fixed $\bar{a}$, region (b) shrinks for a positive cosmological constant $\bar{\Lambda} > 0$ and grows if $\bar{\Lambda} < 0$, what can be seen from the expression for the boundaries of (b), $\bar{L}_z = \bar{a}E \pm \frac{\sqrt{\bar{K}}}{\chi}$. The dependence of region (d) on the parameters $\bar{K}$, $\bar{a}$ and $\bar{\Lambda}$ is much more involved. The upper boundary of (d) is also the lower boundary of region (b). The lower boundary of (d) is in the case of $\epsilon = 0$ given by (4.3.17), but for $\epsilon = 1$ in a complicated parametric form which makes it apparently impossible to determine an explicit connection between the form of region (d) and the parameters. However, the point where the upper and lower boundaries of region (d) touch each other

4. Geodesics in axially symmetric space-times

is where 0 is a double zero of Θ_ν, which is given by $x = 0$ in $E(x)$ and $\bar{L}_z(x)$ from (4.3.16),

$$E(0) = \frac{1}{2}\frac{\bar{a}^2 + \bar{K}(1-\bar{a}^2\bar{\Lambda})}{\bar{a}\sqrt{\bar{K}\chi}}, \quad \bar{L}_z(0) = \frac{1}{2}\frac{\bar{a}^2 - \bar{K}\chi}{\sqrt{\bar{K}\chi}} \quad \text{for} \quad \epsilon = 1,$$

$$E(0) = \frac{1}{2}\frac{\sqrt{\bar{K}}(1-\bar{a}^2\bar{\Lambda})}{\bar{a}\chi}, \quad \bar{L}_z(0) = -\frac{\sqrt{\bar{K}}}{2} \quad \text{for} \quad \epsilon = 0.$$

(4.3.19)

The regions (b) and (d) can be characterized in terms of the modified Carter constant $\mathcal{C}$ in the same way as for $\bar{\Lambda} = 0$. As in region (d) $\Theta_\nu(0) < 0$ it follows that this region corresponds to $\mathcal{C} < 0$. In the same way we can conclude that region (b), where $\Theta_\nu(0) > 0$, corresponds to $\mathcal{C} > 0$.

Types of radial motion

In Sec. 4.1 it was explained that the real zeros of

$$\bar{R}_{\text{KdS}} = \chi^2((\bar{r}^2 + \bar{a}^2)E - \bar{a}L_z)^2 - \Delta_{\bar{r},\text{KdS}}(\epsilon\bar{r}^2 + \bar{K})$$

are extremal values of $\bar{r}(\gamma)$ and, thus, that their number together with the sign of $\bar{R}_{\text{KdS}}$ between them determine the type of geodesic. The polynomial $\bar{R}_{\text{KdS}}$ is in general of degree six in $\bar{r}$ and, therefore, has six possibly complex zeros of which the real zeros are of interest for the type of motion. As in Kerr space-times negative values of $\bar{r}$ are also possible.

The number of real zeros, which determines the type of orbit, changes at that combination of parameters $\bar{a}$, $\bar{\Lambda}$, ϵ, E, $\bar{L}_z$, and K for which two zeros of $\bar{R}_{\text{KdS}}$ coincide. For fixed $\bar{a}$, $\bar{\Lambda}$, ϵ, and K the parameters E and $\bar{L}_z$ for which this happens can be derived by a comparison of coefficients from

$$\bar{R}_{\text{KdS}}(x) = (\bar{r} - x)^2(a_4\bar{r}^4 + a_3\bar{r}^3 + a_2\bar{r}^2 + a_1\bar{r} + a_0),$$

(4.3.20)

where a_i are some real constants with $a_4 = a_3 = 0$ in the case of $\epsilon = 0$. This yields two expressions for $E(x)$ and $\bar{L}_z(x)$, where x is the position of the double zero, which are even more complicated as in the case of $\bar{\Lambda} = 0$. A typical result in slow Kerr-de Sitter including the results of the foregoing subsection is shown in Fig. 4.11 for timelike geodesics and $\Lambda < 0$ and in Figs. 4.12 and 4.13 for small $\Lambda > 0$. As in Kerr space-time, the boundary of region (I) correspond to negative double zeros x, other solid lines to positive x, and dashed lines to regions of θ motion. The alterations of regions of timelike geodesics for varying $\bar{a}^2 < 1$ and $\bar{K}$ are analogous to vanishing $\bar{\Lambda}$, see Fig. 4.5.

In the following, we first shortly discuss the resulting types of orbits for timelike geodesics in Kerr-anti-de Sitter space-time ($\Lambda < 0$) and proceed with an analysis of types of orbits for null as well as timelike geodesics in Kerr-de Sitter space-time ($\Lambda > 0$). In the latter case also exceptional orbits will be discussed.

4.3. Kerr-de Sitter space-time

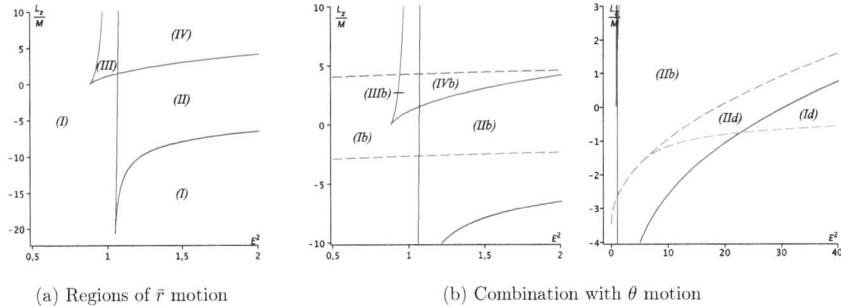

Figure 4.11: Regions of different types of timelike geodesics in Kerr-anti-de Sitter space-time for both $\bar{r}$ and θ motion. Here $\bar{a} = 0.8$, $\bar{\Lambda} = -10^{-5}$, and $\bar{K} = 12$. Solid lines correspond to boundaries of the $\bar{r}$ motion and dashed lines to boundaries of the θ motion.

Case $\Lambda < 0$. In the case of timelike geodesics and a negative cosmological constant region (V) from $\Lambda = 0$ merges with region (I) and region (III) becomes larger for $\Lambda < 0$ due to the shift of the $E^2 = 1$ line to the right. Compared to the situation for $\Lambda = 0$ the possible orbit types in region (III) do not change, but a set of parameters located there may be located in a different region for $\Lambda = 0$. Let us examine the possible orbit types in the remaining regions (I), (II), and (IV).

(I) Here $\bar{R}_{\text{KdS}}$ has two real zeros $r_1 < r_2$ and $\bar{R}_{\text{KdS}}(\bar{r}) \geq 0$ for $r_1 \leq \bar{r} \leq r_2$. Possible orbit types: bound orbit.

(II) (and (III)): $\bar{R}_{\text{KdS}}$ has four real zeros with $\bar{R}_{\text{KdS}}(\bar{r}) \geq 0$ for $r_{2k-1} \leq \bar{r} \leq r_{2k}$, $k = 1, 2$. Possible orbit types: two different bound orbits.

(IV) In this case all six zeros of $\bar{R}_{\text{KdS}}$ are real and $\bar{R}_{\text{KdS}}(\bar{r}) \geq 0$ for $r_{2k-1} \leq \bar{r} \leq r_{2k}$, $k = 1, 2, 3$. Possible orbit types: three different bound orbits.

Concerning crossover orbits, region (III) and (apparently) region (IV) again only contain region (b) of θ motion. Also region (II) intersects both (b) and (d) whereas region (I) can only contain region (d) of θ motion.

Summarizing, the types of orbits significantly change if $E^2 > 1$. The transit orbit in region (I) for $\Lambda = 0$ is transformed to a bound orbit for $\Lambda < 0$ as well as the flyby orbits in regions (II) and (IV). Although region (V) for $\Lambda = 0$ merges with region (I) for $\Lambda < 0$, the types of orbits do not change there. In general, because of $R \to -\infty$ if $\bar{r} \to \pm\infty$ we can not have orbits reaching $\bar{r} = \pm\infty$ at all as expected due to the attractive cosmological force related to $\Lambda < 0$.

All orbit types for $\Lambda < 0$ are summarized in Tab. 4.2

4. Geodesics in axially symmetric space-times

region	−	+	range of $\bar{r}$	types of orbits
Id	1	1	●——●—→	crossover bound
IIb	2	2	●—●—+—●—●—→	2x bound
IId	3	1	●—●—●—+—●—→	bound, crossover bound
IIIb	0	4	+—●—●—●—●—→	2x bound
IVb	2	4	●—●—+—●—●—●—●—→	3x bound

Table 4.2: Orbit types for $\Lambda < 0$. For the description of the +, − and range of $\bar{r}$ columns see Tab. 4.1.

Case $\Lambda > 0$. Let us analyse now which regions change for a positive cosmological constant if compared to the case of $\Lambda = 0$. For null geodesics the possible orbit types in regions (I), (II), and (IV) are the same as for $\Lambda = 0$ but the boundaries are slightly deformed. For timelike geodesics region (V) of $\Lambda = 0$ merged with region (IV), and region (III) becomes smaller for $\Lambda > 0$ due to the shift of the separating $E^2 = 1$ line towards the left. A comparison of the possible orbit types for $\Lambda > 0$ with the one for $\Lambda = 0$ shows that in regions (I) and (II) there are no differences. However, these regions are slightly deformed (for small Λ) and a pair of parameters $(E^2, \bar{L}_z)$ located in region (I) or (II) for $\Lambda > 0$ may be located in a different region for $\Lambda = 0$. For convenience, the possible orbit types for regions (I) and (II) are listed below together with the changed regions (III) and (IV). (Here again we assume $r_i < r_{i+1}$.)

(I) Here all zeros of $\bar{R}_{\text{KdS}}$ are complex and $\bar{R}_{\text{KdS}}(\bar{r}) \geq 0$ for all $\bar{r}$. Possible orbit types: transit orbit.

(II) $\bar{R}_{\text{KdS}}$ has two real zeros r_1, r_2 and $\bar{R}_{\text{KdS}}(\bar{r}) \geq 0$ for $\bar{r} \leq r_1$ and $r_2 \leq \bar{r}$. Possible orbit types: two flyby orbits, one to $+\infty$ and one to $-\infty$.

(III) All six zeros r_i of $\bar{R}_{\text{KdS}}$ are real and $\bar{R}_{\text{KdS}}(\bar{r}) \geq 0$ for $\bar{r} \leq r_1, r_6 \leq \bar{r}$ and $r_{2k} \leq \bar{r} \leq r_{2k+1}$ for $k = 1, 2$. Possible orbit types: two flyby orbits, one to each of $\pm\infty$, and two different bound orbits.

(IV) $\bar{R}_{\text{KdS}}$ has four real zeros and $\bar{R}_{\text{KdS}}(\bar{r}) \geq 0$ for $\bar{r} \leq r_1, r_2 \leq \bar{r} \leq r_3, r_4 \leq \bar{r}$. Possible orbit types: two flyby orbits, one to each of $\pm\infty$ and a bound orbit.

For each region, corresponding effective potentials are shown in Fig. 4.14. Analogous to $\Lambda = 0$, regions (III) and (apparently) (IV) only contain region (b) of the θ motion implying that there are no crossover orbits. Region (I) can only intersect region (d) because only transit orbits are possible. The remaining region (II) is the only one proven to intersect regions (b) and (d). In Fig. 4.15 some

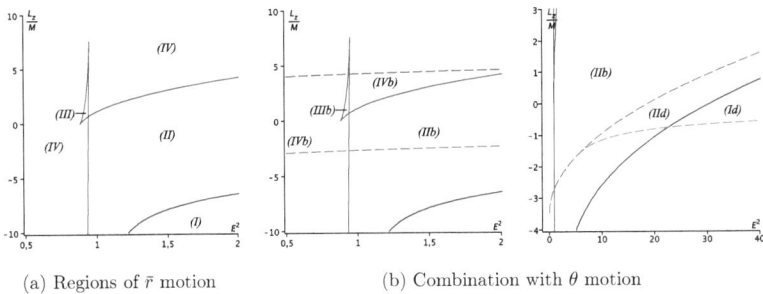

Figure 4.12: Regions of different types of timelike geodesics in Kerr-de Sitter space-time for both $\bar{r}$ and θ motion. Here $\bar{a} = 0.8$, $\bar{\Lambda} = 10^{-5}$, and $\bar{K} = 12$. Solid lines correspond to boundaries of the $\bar{r}$ motion and dashed lines to boundaries of the θ motion.

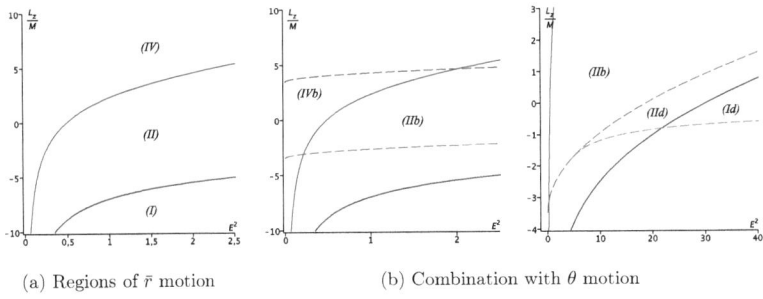

Figure 4.13: Regions of different types of null geodesics in Kerr-de Sitter space-time for both $\bar{r}$ and θ motion. Here $\bar{a} = 0.8$, $\bar{\Lambda} = 10^{-5}$, and $\bar{K} = 12$. Solid lines correspond to boundaries of the $\bar{r}$ motion and dashed lines to boundaries of the θ motion.

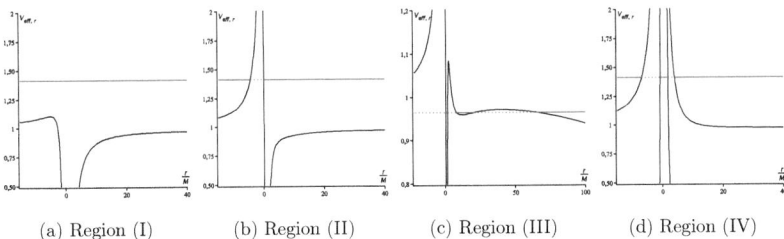

Figure 4.14: Effective potentials for different regions of the $\bar{r}$ motion in Kerr-de Sitter space-time. The horizontal lines denote the energy parameter E.

4. Geodesics in axially symmetric space-times

region	−	+	range of $\bar{r}$	types of orbits
Id	0	0	———|———	transit
IIb	1	1	—•——|——•—	2x flyby
IId	2	0	—•—•—|———	flyby, crossover flyby
IIIb	1	5	—•———|—•—•—•—•—	2x flyby, 2x bound
IVb	1	3	—•———|—•—•—•—	2x flyby, bound

Table 4.3: Orbit types of timelike geodesics in Kerr-de Sitter space-time for small $\Lambda > 0$. For the description of the $+$, $-$ and range of $\bar{r}$ columns see Tab. 4.1.

typical orbits for different regions are shown, which were created using the analytical solution of the geodesic equation presented in the next section.

We conclude that for $E^2 > 1$ the types of timelike orbits are not noticeably changed, whereas for $E^2 \leq 1$ there are significant changes. In the former region (V) (for $\Lambda = 0$), which is now in region (IV), and in region (III) we have two additional flyby orbits which are not present for $\Lambda = 0$. In a small vertical stripe left of $E^2 = 1$ there are even orbits which are bound for $\Lambda = 0$ but reaching infinity for $\Lambda > 0$. In particular, it is independent of the value of E if a geodesic may reach infinity as expected from the repulsive cosmological force related to $\Lambda > 0$.

Note that for huge Λ the separation in regions (I) to (IV) as explained above is no longer possible because the repulsive cosmological force becomes so strong that all bound orbits become flyby orbits. In this case region (III) vanishes and we have only two regions, one with two real zeros corresponding to two flyby orbits and one with only complex zeros corresponding to a transit orbit.

Examples of orbits which highlight the influence of Λ on the geodesics are illustrated in Fig. 4.16. From the discussion in this section it can be inferred that for $\Lambda > 0$ there are four parameter regions where the changes compared to $\Lambda = 0$ are most obvious. The first two are the regions (III) and (IV) with $E^2 < 1$, where we have additional flyby orbits not present for $\Lambda = 0$. Third and fourth, the shift from region (V) of $\Lambda = 0$ to region (II) of $\Lambda > 0$ for $E^2 = 1 - \delta$, $\delta > 0$ small, and the shift from region (III) of $\Lambda = 0$ to region (IV) of $\Lambda > 0$, again for $E^2 = 1 - \delta$ are most interesting as the (outer) bound orbit becomes a flyby orbit.

All orbit types for small $\Lambda > 0$ are summarized in Tab. 4.3.

Exceptional orbits

In this section some exceptional orbits related to multiple zeros of $\bar{R}_{\text{KdS}}$, i.e. spherical orbits with constant r and orbits asymptotically approaching a constant r, will be discussed. There are two

4.3. Kerr-de Sitter space-time

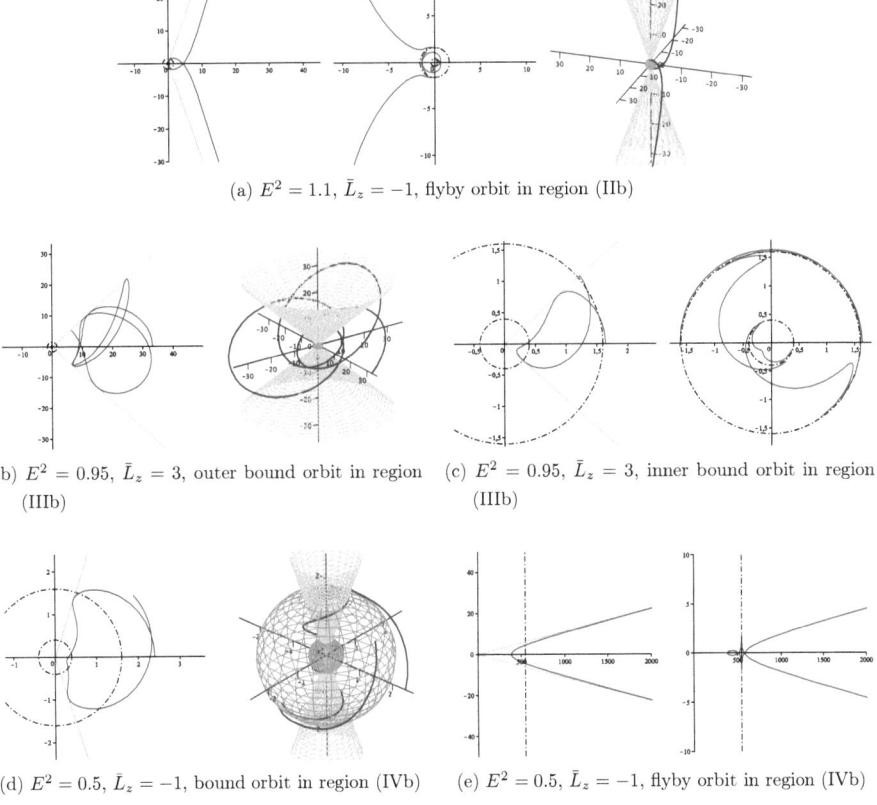

(a) $E^2 = 1.1$, $\bar{L}_z = -1$, flyby orbit in region (IIb)

(b) $E^2 = 0.95$, $\bar{L}_z = 3$, outer bound orbit in region (IIIb)

(c) $E^2 = 0.95$, $\bar{L}_z = 3$, inner bound orbit in region (IIIb)

(d) $E^2 = 0.5$, $\bar{L}_z = -1$, bound orbit in region (IVb)

(e) $E^2 = 0.5$, $\bar{L}_z = -1$, flyby orbit in region (IVb)

Figure 4.15: Timelike geodesics in the Kerr-de Sitter space-time with $\bar{a} = 0.8$, $\bar{\Lambda} = \frac{1}{3}10^{-5}$, and $\bar{K} = 12$. In each subplot, on the left side the r-θ plane and on the right side a 3d-image and/or the r-φ plane is shown. Light grey lines and cones correspond to extremal θ and dark grey spheres as well as dashed black circles to horizons. The bound orbits in (c) and (d) cross the Cauchy horizon several times and, thus, are many world bound orbits. The flyby orbit in (a) is a two-world orbit.

4. Geodesics in axially symmetric space-times

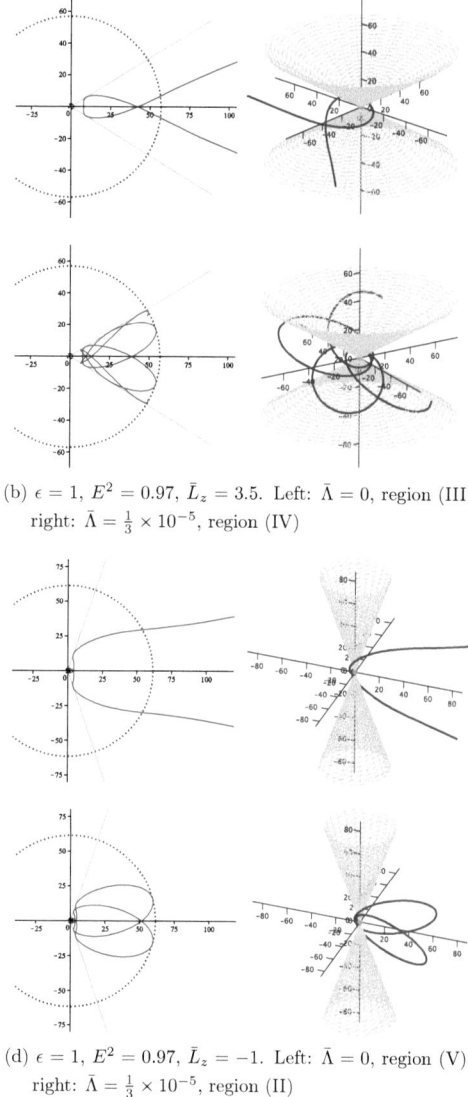

(b) $\epsilon = 1$, $E^2 = 0.97$, $\bar{L}_z = 3.5$. Left: $\bar{\Lambda} = 0$, region (III), right: $\bar{\Lambda} = \frac{1}{3} \times 10^{-5}$, region (IV)

(d) $\epsilon = 1$, $E^2 = 0.97$, $\bar{L}_z = -1$. Left: $\bar{\Lambda} = 0$, region (V), right: $\bar{\Lambda} = \frac{1}{3} \times 10^{-5}$, region (II)

Figure 4.16: Comparison between orbits in Kerr and Kerr-de Sitter space-time for $\bar{a} = 0.8$ and $\bar{K} = 12$. In (a) the maximal r in Kerr space-time is $\bar{r}_{\max} \approx 57.01$ and in (b) $\bar{r}_{\max} \approx 61.56$ indicated by the dotted lines.

4.3. Kerr-de Sitter space-time

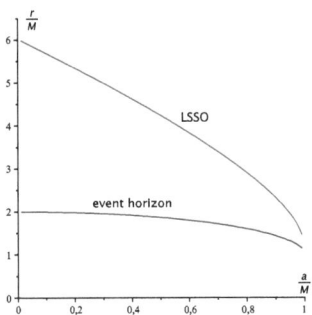

Figure 4.17: Radius of last stable spherical orbit and event horizon for $\bar{\Lambda} = 10^{-5}$ and varying $\bar{a}$. Energy E, angular momentum $\bar{L}_z$, and Carter constant K are free parameters.

types of spherical orbits: stable and unstable. Stable spherical orbits with $\bar{r}(\gamma) \equiv \bar{r}_0$ occur if radial coordinates adjacent to $\bar{r}_0$ are not allowed due to $\bar{R}_{\text{KdS}}(\bar{r}) < 0$, which happens if $\bar{r}_0$ is a maximum of $\bar{R}_{\text{KdS}}$. Unstable spherical orbits with $\bar{r}(\gamma) \equiv \bar{r}_0$ are trajectories where radial coordinates $\bar{r}$ in the neighborhood of $\bar{r}_0$ with $\bar{r} < \bar{r}_0$ or $\bar{r} > \bar{r}_0$ are allowed. Therefore, these orbits are related to a minimum or to an inflection point of $\bar{R}_{\text{KdS}}$. If $\bar{r}_0$ is an inflection point, an asymptotic approach to $\bar{r}_0$ is only possible from one side of $\bar{r}_0$ whereas this is possible from both sides if $\bar{r}_0$ is a minimum of $\bar{R}_{\text{KdS}}$. Asymptotic orbits can also be divided into two types: unbound and bound. The latter case corresponds to orbits which approach for both $t \to \infty$ and $t \to -\infty$ a spherical orbit. Bound asymptotic orbits are also known as homoclinic orbits. If the asymptotic orbit is unbound it reaches $\bar{r} = \infty$ for either $t \to \infty$ or $t \to -\infty$.

For asymptotic bound or unbound trajectories corresponding to an unstable spherical orbit the equations of motion simplify considerably. In this case the equation for $\bar{r}(\gamma)$ as well as the $\bar{r}$ dependent integrals in the φ and t equations are of elliptic type and can be solved in terms of Weierstrass elliptic functions, see (4.3.34), (4.3.47), and (4.3.59). Note that these solutions are not limited to the case of equatorial circular orbits but are valid for all types of asymptotic orbits and, thus, generalize the analytical solutions for homoclinic orbits in [76] not only to Kerr-de Sitter space-time but also to arbitrary inclinations.

From all spherical orbits the Last Stable Spherical Orbit (LSSO) and, in particular, the Innermost Stable Circular Orbit (ISCO) in the equatorial plane are of importance as they represent the transition from stable orbits to those which fall through the event horizon. The corresponding multiple zero of $\bar{R}_{\text{KdS}}$ appears at the boundaries of the different regions of r motion, see Fig. 4.12.

4. Geodesics in axially symmetric space-times

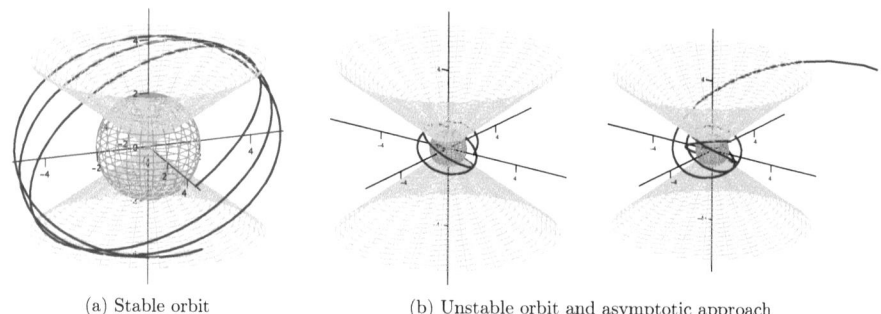

(a) Stable orbit

(b) Unstable orbit and asymptotic approach

Figure 4.18: (a) Last Stable Spherical orbit at $\bar{r}(\gamma) \equiv 4.864$ for $\bar{a} = 0.4$, $\bar{\Lambda} = 10^{-5}$, and $\bar{K} = 8$. The corresponding parameter values are (approximately) $E^2 = 0.86238633$ and $\bar{L}_z = 2.5063691$. (b) Unstable spherical orbit with $\bar{r}(\gamma) \equiv 1.5$ (left) and asymptotic approach (right). Here $\bar{a} = 0.6$, $\bar{\Lambda} = 10^{-5}$, $\bar{K} = 8$, $E = 0.9720311146$ and $\bar{L}_z = 5.1355914740$ (boundary of region (IIIb)).

For given $\bar{a}$, $\bar{\Lambda}$, and K the LSSO can be determined from

$$\bar{R}_{\text{KdS}}(\bar{r}) = 0, \quad \frac{d\bar{R}_{\text{KdS}}}{d\bar{r}}(\bar{r}) = 0, \quad \text{and} \quad \frac{d^2\bar{R}_{\text{KdS}}}{d\bar{r}^2}(\bar{r}) = 0 \quad (4.3.21)$$

for $\bar{r} \geq \bar{r}_h$ with the event horizon $\bar{r}_h$. Note that in the case of the ISCO it is necessarily $\bar{K} = \chi^2(\bar{a}E - \bar{L}_z)^2$ (condition for equatorial orbits) and, thus, only $\bar{a}$ and $\bar{\Lambda}$ have to be known for a calculation. The solutions of (4.3.21) are limiting cases of the stable spherical (or, for $\bar{K} = \chi^2(\bar{a}E - \bar{L}_z)^2$, circular) orbits, and are given by the corner points on the boundaries of region (III) of the $\bar{r}$ motion. From the results of (4.3.21) we search for the smallest possible double zero $\bar{r}$ which is a maximum. In general, this will be the lower left corner of the boundary of region (III). In the case of the ISCO in the equatorial plane we are now done. For the LSSO, we have to check in addition whether the corresponding values of $E^2(\bar{r})$ and $\bar{L}_z(\bar{r})$ (given by (4.3.20)) are located in an allowed region of the θ motion. If this is the case, we found the LSSO. If not, we can determine the LSSO as the intersection point of the boundary of region (III) with a boundary of an allowed θ region. Note that it is not possible to determine an LSSO (for given $\bar{a}$, $\bar{\Lambda}$, and $\bar{K}$) if there is no spherical orbit at all outside the event horizon which happens if no boundary of the $\bar{r}$ motion is located in an allowed region of the θ motion. As an example, this is the case for $\bar{\Lambda} = 10^{-5}$, $\bar{a} = 0.2$, and $\bar{K} = 0.4$. Also, the LSSO is identical with the ISCO if it is given as an intersection point with the boundary of region (b) of the θ motion.

If K is left as a free parameter and only $\bar{a}$ and $\bar{\Lambda}$ are fixed, the LSSO can be found by a minimization of $\bar{r}$ under the constraints

$$\bar{r}_h \leq \bar{r} \leq \bar{r}_c, \quad E^2(\bar{r}) \geq 0, \quad K \geq 0 \tag{4.3.22}$$

$$\frac{d\bar{R}_{\text{KdS}}}{d\bar{r}}(\bar{r}) \leq 0, \tag{4.3.23}$$

$$\bar{L}_z(\bar{r}) \leq \bar{a}E + \frac{\sqrt{K}}{\chi}, \quad \bar{L}_z(\bar{r}) \geq \bar{a}E - \frac{\sqrt{K}}{\chi}, \tag{4.3.24}$$

with $E^2(\bar{r})$ and $\bar{L}_z(\bar{r})$ as derived from (4.3.20). Here Eq. (4.3.22) corresponds to general constraints, i.e. the LSSO should be located between the event horizon $\bar{r}_h$ and the cosmological horizon $\bar{r}_c$, the energy parameter should have a real value, and $K > 0$ by theorem 4.2. Eq. (4.3.23) ensures that $\bar{r}$ corresponds to a stable orbit, Eq. (4.3.24) that the LSSO lies in a region where θ motion is allowed. For different values of $\bar{a}$ and $\bar{\Lambda} = 10^{-5}$ the results of this analysis are shown in Fig. 4.17. For examples of spherical orbits see Fig. 4.18.

4.3.2 Analytical solution of geodesic equations

We will now analytically solve the geodesic equation in Kerr-de Sitter space-time (4.3.8) - (4.3.11)

$$\left(\frac{d\bar{r}}{d\gamma}\right)^2 = \bar{R}_{\text{KdS}}(\bar{r}) = \chi^2 \mathbb{P}^2(r) - \Delta_{\bar{r},\text{KdS}}(\epsilon \bar{r}^2 + \bar{K}), \tag{4.3.8}$$

$$\left(\frac{d\theta}{d\gamma}\right)^2 = \bar{\Theta}_{\text{KdS}}(\theta) = \Delta_\theta(\bar{K} - \epsilon\bar{a}^2 \cos^2\theta) - \frac{\chi^2 \mathbb{T}^2(\theta)}{\sin^2\theta}, \tag{4.3.9}$$

$$\frac{d\varphi}{d\gamma} = \frac{\bar{a}}{\Delta_{\bar{r},\text{KdS}}}\mathbb{P}(r) - \frac{1}{\Delta_\theta \sin^2\theta}\mathbb{T}(\theta), \tag{4.3.10}$$

$$\frac{d\bar{t}}{d\gamma} = \frac{\bar{r}^2 + \bar{a}^2}{\Delta_{\bar{r},\text{KdS}}}\mathbb{P}(r) - \frac{\bar{a}}{\Delta_\theta}\mathbb{T}(\theta), \tag{4.3.11}$$

where $\mathbb{P}$ and $\mathbb{T}$ are given by

$$\mathbb{P}(r) = (\bar{r}^2 + \bar{a}^2)E - \bar{a}\bar{L}_z,$$

$$\mathbb{T}(\theta) = \bar{a}E \sin^2\theta - \bar{L}_z.$$

Each equation will be treated separately.

θ motion

We begin with the differential equation (4.3.9)

$$\left(\frac{d\theta}{d\gamma}\right)^2 = \bar{\Theta}_{\text{KdS}}(\theta) = \Delta_\theta(\bar{K} - \epsilon\bar{a}^2 \cos^2\theta) - \frac{\chi^2 \mathbb{T}^2(\theta)}{\sin^2\theta},$$

4. Geodesics in axially symmetric space-times

which can be simplified by the substitution $\nu = \cos^2\theta$ yielding

$$\left(\frac{d\nu}{d\gamma}\right)^2 = 4\nu\Theta_\nu, \qquad (4.3.25)$$

where Θ_ν is the polynomial of degree 3 defined in (4.3.15). This differential equation can be solved easily if $4\nu\Theta_\nu$ has a zero with multiplicity 2 or more. In this case (4.3.25) can be rewritten as

$$\gamma - \gamma_0 = \int_{\nu_0}^{\nu} \frac{d\nu'}{(\nu' - \nu_i)^j \sqrt{P_2(\nu')}}, \qquad (4.3.26)$$

where γ_0 and ν_0 are initial values, P_2 is a polynomial with maximum degree 2, and ν_i is a zero of $4\nu\Theta_\nu$ with multiplicity $2j$ or $2j+1$, $j = 1, 2$. The integral on the right hand side can then be solved by elementary functions [31]. As in this case the explicit expression provides no further insight and some case distinctions would be necessary we skip the solution procedure.

Timelike geodesics If $4\nu\Theta_\nu$ has only simple zeros the differential equation (4.3.25) is of elliptic type and first kind and can be solved in terms of the Weierstrass elliptic function $\wp$. To obtain a solution we transform $4\nu\Theta_\nu$ to the Weierstrass form $(4y^3 - g_2 y - g_3)$ for some constants g_2 and g_3 by the standard procedure described in (2.1.9): First, we substitute $\nu = \xi^{-1}$ giving

$$\left(\frac{d\xi}{d\gamma}\right)^2 = \Theta_\xi, \qquad (4.3.27)$$

where

$$\Theta_\xi := 4\xi^3\left(\bar{K} - \chi^2(\bar{a}E - \bar{L}_z)^2\right) + 4\xi^2\left(\bar{a}^2(\bar{K}\bar{\Lambda} - \epsilon) - \bar{K} + 2\chi^2\bar{a}(\bar{a}E - \bar{L}_z)\right)$$
$$+ 4\bar{a}^2(\epsilon(1 - \bar{\Lambda}\bar{a}^2) - \chi^2 - \bar{\Lambda}\bar{K})\xi + 4\epsilon\bar{a}^4\bar{\Lambda} \qquad (4.3.28)$$
$$=: \sum_{i=1}^{3} a_i \xi^i .$$

Second, the substitution $\xi = \frac{1}{a_3}\left(4y - \frac{a_2}{3}\right)$ implies

$$\left(\frac{dy}{d\gamma}\right)^2 = 4y^3 - g_2 y - g_3, \qquad (4.3.29)$$

where g_2, g_3 are given by (2.1.11)

$$g_2 = \frac{1}{16}\left(\frac{4}{3}a_2^2 - 4a_1 a_3\right),$$

$$g_3 = \frac{1}{16}\left(\frac{1}{3}a_1 a_2 a_3 - \frac{2}{27}a_2^3 - a_0 a_3^2\right).$$

The differential equation (4.3.29) is of elliptic type and first kind, which can be solved by (2.1.8)

$$y(\gamma) = \wp(\gamma - \gamma_{\theta,\text{in}}; g_2, g_3). \tag{4.3.30}$$

Accordingly, the solution of (4.2.9) is given by

$$\theta(\gamma) = \arccos\left(\pm\sqrt{\frac{a_3}{4\wp(\gamma - \gamma_{\theta,\text{in}}; g_2, g_3) - \frac{a_2}{3}}}\right) \quad \text{for} \quad \epsilon = 1, \tag{4.3.31}$$

where $\gamma_{\theta,\text{in}} = \gamma_0 + \int_{y_0}^{\infty} \frac{dy'}{\sqrt{4y'^3 - g_2 y' - g_3}}$ with $y_0 = \frac{a_3}{4\cos^2(\theta_0)} + \frac{a_2}{12}$ depends on the initial values γ_0 and θ_0 only. The sign of the square root depends on whether $\theta(\gamma)$ should be in $(0, \frac{\pi}{2})$ (positive sign) or in $(\frac{\pi}{2}, \pi)$ (negative sign) and reflects the symmetry of the θ motion with respect to the equatorial plane $\theta = \frac{\pi}{2}$. If the motion is located in region (b) from the previous section this implies that the two solutions have to be glued together along $\theta(\gamma) = \frac{\pi}{2}$ if the whole θ motion should be considered.

Null geodesics In contrast to the $\bar{r}$ motion considered in the next subsection, the structure of (4.3.25) does not considerably simplify if we consider light with $\epsilon = 0$. The only difference to the solution method outlined above is that $4\nu\Theta_\nu$ is already a polynomial of degree 3 and, thus, that the substitution $\nu = \xi^{-1}$ is not necessary. Indeed, the standard substitution $\nu = \frac{1}{b_3}\left(4y - \frac{b_2}{3}\right)$ where $4\nu\Theta_\nu = \sum_{i=1}^{3} b_i \nu^i$ transforms the problem to the form (4.3.29). The solution is then given by

$$\theta(\gamma) = \arccos\left(\pm\sqrt{\frac{4}{b_3}\wp(\gamma - \gamma_{\theta,\text{in}}; g_2, g_3) - \frac{b_2}{3b_3}}\right) \quad \text{for} \quad \epsilon = 0, \tag{4.3.32}$$

where $\gamma_{\theta,\text{in}}$, g_2, and g_3 are as above with a_i replaced by b_i.

r motion

The differential equation that describes the dynamics of r

$$\left(\frac{d\bar{r}}{d\gamma}\right)^2 = \bar{R}_{\text{KdS}}(\bar{r}) = \chi^2 \mathbb{P}^2(r) - \Delta_{\bar{r},\text{KdS}}(\epsilon\bar{r}^2 + \bar{K}) \tag{4.3.8}$$

is more complicated because $\bar{R}_{\text{KdS}}$ is a polynomial of a degree up to 6. If $\bar{R}_{\text{KdS}}$ has a zero of multiplicity 4 or more, two zeros of multiplicity 2 or more, or if $\epsilon = 0$ and $\bar{R}_{\text{KdS}}$ has a zero of multiplicity 2 or more the differential equation (4.2.8) can be written as

$$\gamma - \gamma_0 = \int_{\bar{r}_0}^{\bar{r}} \frac{d\bar{r}'}{\prod_{i=1}^{k}(\bar{r}' - \bar{r}_i)^{j_i} \sqrt{P_2(\bar{r}')}}, \tag{4.3.33}$$

where γ_0 and $\bar{r}_0$ are initial values, P_2 is a polynomial with maximum degree 2, $\bar{r}_i$ are zeros of $\bar{R}_{\text{KdS}}$ with multiplicity $2j_i$ or $2j_i + 1$ where $j_i = 1, 2$, and $k = 2$ if there are two zeros of multiplicity 2 or more and $k = 1$ else. The integral on the right hand side can then be solved by elementary functions [31]. As the explicit expression provides no further insight and some case distinctions would be necessary we skip the solution procedure.

4. Geodesics in axially symmetric space-times

Null geodesics If we consider light, i.e. $\epsilon = 0$, $\bar{R}_{\text{KdS}}$ is in general of degree 4 and the differential equation (4.3.8) is of elliptic type and first kind. Thus, (4.3.8) can be solved using the standard method for this type of equation, see (2.1.5): With the substitutions $\bar{r} = \xi^{-1} + \bar{r}_{\text{KdS}}$, where $\bar{r}_{\text{KdS}}$ is a zero of $\bar{R}_{\text{KdS}}$, and $\xi = \frac{1}{b_3}\left(4y - \frac{b_2}{3}\right)$, where $b_i = \frac{1}{(4-i)!}\frac{d^{(4-i)}\bar{R}_{\text{KdS}}}{d\bar{r}^{(4-i)}}(\bar{r}_{\text{KdS}})$, we arrive at the standard Weierstrass form (2.1.5). The solution can then be formulated in terms of the Weierstrass elliptic functions. The result is

$$\bar{r}(\gamma) = \frac{b_3}{4\wp(\gamma - \gamma_{\bar{r},\text{in}}; g_2, g_3) - \frac{b_2}{3}} + \bar{r}_{\text{KdS}}, \tag{4.3.34}$$

where $\gamma_{\bar{r},\text{in}} = \gamma_0 + \int_{y_0}^{\infty} \frac{dy'}{\sqrt{4y'^3 - g_2, r y' - g_3, r}}$ and $y_0 = \frac{b_3}{4(\bar{r}_0 - r_{\text{R}})} + \frac{b_2}{12}$ depend only on the initial values $\gamma_0, \bar{r}_0$, and g_2, g_3 are defined in (2.1.11).

Timelike geodesics If we consider particles, i.e. $\epsilon = 1$, the differential equation (4.3.8) is also of elliptic type but of the third kind if $\bar{R}_{\text{KdS}}$ has a double or triple zero $\bar{r}_1$. In this case (4.2.8) reads

$$\gamma - \gamma_0 = \int_{\bar{r}_0}^{\bar{r}} \frac{d\bar{r}'}{(r - r_1)\sqrt{P_4(\bar{r})}}, \tag{4.3.35}$$

where P_4 is a polynomial of degree 4. This equation can be solved for $\bar{r}(\gamma)$ with the method presented in Thm. 2.5 and appendix A.

If we assume that $\bar{R}_{\text{KdS}}$ has only simple zeros the differential equation (4.3.8) is of hyperelliptic type. It can be solved in terms of derivatives of the Kleinian σ function with the method already used for the solution of the geodesic equation in Schwarzschild-de Sitter space-time, see Chap. 3, Sec. 3.3. For this, we have to cast (4.3.8) into the standard form by a substitution $\bar{r} = \pm\frac{1}{u} + \bar{r}_{\text{KdS}}$ with a zero $\bar{r}_{\text{KdS}}$ of $\bar{R}_{\text{KdS}}$. This yields

$$\left(u\frac{du}{d\gamma}\right)^2 = c_5 R_u, \tag{4.3.36}$$

where

$$R_u = \sum_{i=0}^{5} \frac{c_i}{c_5} u^i, \quad c_i = \frac{(\pm 1)^i}{(6-i)!}\frac{d^{(6-i)}\bar{R}_{\text{KdS}}}{du^{(6-i)}}(\bar{r}_{\text{KdS}}). \tag{4.3.37}$$

For convenience, the sign in the substitution should be chosen such that the constant c_5 is positive and, therefore, depends on the choice of $\bar{r}_{\text{KdS}}$ and the sign of $\bar{\Lambda}$. The differential equation (4.3.36) is of first kind and can be solved by

$$u(\gamma) = -\frac{\sigma_1}{\sigma_2}\left(\frac{f(\sqrt{c_5}\gamma - \gamma_{\bar{r},\text{in}})}{\sqrt{c_5}\gamma - \gamma_{\bar{r},\text{in}}}\right), \tag{4.3.38}$$

where $\gamma_{\bar{r},\text{in}} = \sqrt{c_5}\gamma_0 + \int_{u_0}^{\infty} \frac{u\,du}{\sqrt{\tilde{R}_u}}$ and $u_0 = \pm(\bar{r}_0 - \bar{r}_{\text{KdS}})^{-1}$ depends only on the initial values γ_0 and $\bar{r}_0$. Here f is the function that describes the θ-divisor, i.e. $\sigma((f(x),x)^t) = 0$, see Chap. 3. The radial distance $\bar{r}$ is then given by

$$\bar{r}(\gamma) = \mp\frac{\sigma_2}{\sigma_1}\left(\frac{f(\sqrt{c_5}\gamma - \gamma_{\bar{r},\text{in}})}{\sqrt{c_5}\gamma - \gamma_{\bar{r},\text{in}}}\right) + \bar{r}_{\text{KdS}}, \qquad (4.3.39)$$

where the sign depends on the sign chosen in the substitution $\bar{r} = \pm\frac{1}{u} + \bar{r}_{\text{KdS}}$, i.e. is such that c_5 in (4.3.37) is positive.

φ motion

We treat now the most complicated equation of motion in Kerr-de Sitter space-time, namely the equation for the azimuthal angle (4.3.10)

$$\frac{1}{\chi^2}\frac{d\varphi}{d\gamma} = \frac{\bar{a}}{\Delta_{\bar{r},\text{KdS}}}\mathbb{P}(r) - \frac{1}{\Delta_\theta \sin^2\theta}\mathbb{T}(\theta). \qquad (4.3.10)$$

This equation can be splitted in a part only dependent on $\bar{r}$ and in a part only dependent on θ. Integration yields

$$\begin{aligned}
\varphi - \varphi_0 &= \chi^2\left[\int_{\gamma_0}^{\gamma}\frac{\bar{a}\mathbb{P}(r)}{\Delta_{\bar{r}(\gamma)}}d\gamma - \int_{\gamma_0}^{\gamma}\frac{\mathbb{T}(\theta)d\gamma}{\Delta_{\theta(\gamma)}\sin^2\theta(\gamma)}\right] \\
&= \chi^2\left[\int_{\bar{r}_0}^{\bar{r}}\frac{\bar{a}\mathbb{P}(r)d\bar{r}'}{\Delta_{\bar{r}'}\sqrt{R}} - \int_{\theta_0}^{\theta}\frac{\mathbb{T}(\theta')d\theta'}{\Delta_{\theta'}\sin^2\theta'\sqrt{\Theta}}\right],
\end{aligned} \qquad (4.3.40)$$

where we substituted $\bar{r} = \bar{r}(\gamma)$, i.e. $\frac{d\bar{r}}{d\gamma} = \sqrt{R_{\text{KdS}}}$, in the first and $\theta = \theta(\gamma)$, i.e. $\frac{d\theta}{d\gamma} = \sqrt{\Theta_{\text{KdS}}}$, in the second integral.

We will solve now the two integrals in (4.3.40) separately.

The θ dependent integral Let us consider the integral

$$I_\theta := \int_{\theta_0}^{\theta}\frac{(\sin^2\theta\,\bar{a}E - \bar{L}_z)\,d\theta}{\Delta_\theta\sin^2\theta\sqrt{\bar{\Theta}_{\text{KdS}}}}, \qquad (4.3.41)$$

which can be transformed to the simpler form

$$I_\theta = \mp\int_{\nu_0}^{\nu}\frac{\bar{a}E(1-\nu) - \bar{L}_z}{\Delta_\nu(1-\nu)\sqrt{4\nu\Theta_{\nu'}}}d\nu', \qquad (4.3.42)$$

by the substitution $\nu = \cos^2\theta$, where Θ_ν is defined in (4.3.15) and $\Delta_\nu = 1 + \bar{a}^2\bar{\Lambda}\nu$. As in Kerr space-time, we have to split here the integration part such that every piece is fully contained in the interval

4. Geodesics in axially symmetric space-times

$(0, \frac{\pi}{2}]$ or $[\frac{\pi}{2}, \pi)$. If $\theta \in (0, \frac{\pi}{2}]$ we have $\cos\theta = +\sqrt{\nu}$, but for $\theta \in [\frac{\pi}{2}, \pi)$ it is $\cos\theta = -\sqrt{\nu}$. Thus, for every part of the integration path the sign of the square root of ν has to be chosen appropriately. In the following we assume for simplicity that $\cos\theta = +\sqrt{\nu}$.

Analogous to subsection 4.3.2 the integral I_θ can be solved by elementary functions if $4\nu\Theta_\nu$ has at least a double zero [31]. If $4\nu\Theta_\nu$ has only simple zeros, I_θ is of elliptic type and of third kind. If this is the case, the solution of I_θ is given by

$$I_\theta = \frac{|a_3|}{a_3}\left\{ (\bar{a}E - \bar{L}_z)(v - v_0) - \sum_{i=1}^{4} \frac{a_3}{4\chi\wp'(v_i)}\bigg(\zeta(v_i)(v - v_0) \right.$$
$$\left. + \log\frac{\sigma(v - v_i)}{\sigma(v_0 - v_i)} + 2\pi i k_i \bigg)\left(\bar{a}^3 \bar{\Lambda}(\chi E - \bar{a}\bar{\Lambda}\bar{L}_z)(\delta_{i1} + \delta_{i2}) + \bar{L}_z(\delta_{i3} + \delta_{i4}) \right) \right\} \quad (4.3.43)$$

where the constants a_i are defined as in section 4.3.2,

$$\wp(v_1) = \frac{a_2}{12} - \frac{1}{4}\bar{a}^2\bar{\Lambda}a_3 = \wp(v_2),$$
$$\wp(v_3) = \frac{a_2}{12} + \frac{a_3}{4} = \wp(v_4), \quad (4.3.44)$$

$v = v(\gamma) = \gamma - \gamma_{\theta,\text{in}}$ with $\gamma_{\theta,\text{in}}$ as in (4.3.30) and $v_0 = v(\gamma_0)$. The integers k_i correspond to different branches of log. The details of the computation can be found in appendix A.

The r dependent integral We solve now the $\bar{r}$ dependent integral in (4.3.40)

$$I_r := \int_{\bar{r}_0}^{\bar{r}} \frac{\bar{a}\left((\bar{r}^2 + \bar{a}^2)E - \bar{a}\bar{L}_z\right) d\bar{r}}{\Delta_{\bar{r},\text{KdS}}\sqrt{\bar{R}_{\text{KdS}}}}. \quad (4.3.45)$$

Analogous to subsection 4.3.2 this integral can be solved by elementary functions if $\bar{R}_{\text{KdS}}$ has a zero with multiplicity 4 or more or two zeros with multiplicity 2 or more [31].

Null geodesics For $\epsilon = 0$ the polynomial $\bar{R}_{\text{KdS}}$ is in general of degree 4 and I_r is of elliptic type and third kind. In this case it can be solved analogously to I_θ. The same substitutions $\bar{r} = \frac{1}{\xi} + \bar{r}_{\text{KdS}}$ and $\xi = \frac{1}{b_3}\left(4y - \frac{b_2}{3}\right)$ as in subsection 4.3.2 for the case $\epsilon = 0$, a subsequent partial fraction decomposition, and the final substitution $y = \wp(v)$ result in

$$\frac{b_3}{|b_3|}I_r = \sum_{i=1}^{4} C_i \int_{v_0}^{v} \frac{dv}{\wp(v) - y_i} - \frac{\bar{a}\left(\bar{r}_{\text{KdS}}^2 + \bar{a}^2 - \bar{a}D\right)}{\Delta_{\bar{r}=\bar{r}_{\text{KdS}}}} \int_{v_0}^{v} dv, \quad (4.3.46)$$

where y_i are the four zeros of $\Delta_{y(\bar{r}),\text{KdS}}$, b_3 defined as in (4.3.34), and C_i are the coefficients of the partial fractions dependent on the parameters and $\bar{r}_{\text{KdS}}$. The four functions $f_i(v) = (\wp(v) - y_i)^{-1}$

4.3. Kerr-de Sitter space-time

have simple poles in v_{i1}, v_{i2} with $\wp(v_{i1}) = y_i = \wp(v_{i2})$ and have to be integrated with the method presented in Thm. 2.5 and appendix A. Then I_r is given by

$$\frac{b_3}{|b_3|} I_r = \sum_{i=1}^{4} \sum_{j=1}^{2} \frac{C_i}{\wp'(v_{ij})} \left[\zeta(v_{ij})(v-v_0) + \log \sigma(v-v_{ij}) - \log \sigma(v_0-v_{ij}) \right]$$
$$- \frac{\bar{a}\left((\bar{r}_{\text{KdS}}^2 + \bar{a}^2)E - \bar{a}\bar{L}_z\right)}{\Delta_{\bar{r}=\bar{r}_{\text{KdS}}}} (v-v_0), \quad (4.3.47)$$

where $v = v(\gamma) = \gamma - \gamma_{\bar{r},\text{in}}$, $v_0 = v(\gamma_0)$ with $\gamma_{\bar{r},\text{in}}$ as in (4.3.34). In the same way I_r can be solved if $\epsilon = 1$ and $\bar{R}_{\text{KdS}}$ has a double or triple zero.

Timelike geodesics If we consider particles, i.e. $\epsilon = 1$, and assume that $\bar{R}_{\text{KdS}}$ has only simple zeros, I_r is of hyperelliptic type and third kind. The solution can be found with the help of (2.4.12) and (2.4.15). First, we transform I_r analogously to section 4.3.2 to the standard form by a substitution $\bar{r} = \pm 1/u + \bar{r}_{\text{KdS}}$ with a zero $\bar{r}_{\text{KdS}}$ of $\bar{R}_{\text{KdS}}$. Afterward we simplify the integrand by a partial fraction decomposition which allows us to express I_r in terms of the canonical holomorphic differentials $d\vec{z}$ (2.2.2) and the canonical differential of third kind $dP(x_1, x_2)$ (2.2.10). These differentials can then in turn be expressed in terms of the Mino time γ. The first step results in

$$I_r = \mp \bar{a} \int_{u_0}^{u} \frac{(\pm \frac{1}{u} + \bar{r}_{\text{KdS}})^2 + \bar{a}(\bar{a}E - \bar{L}_z)}{\Delta_{\bar{r}=\pm 1/u + \bar{r}_{\text{KdS}}} \sqrt{u^{-6} c_5 R_u}} \frac{du}{u^2}$$
$$= \mp \bar{a} \int_{u_0}^{u} \frac{[\bar{r}_{\text{KdS}}^2 + \bar{a}(\bar{a}E - \bar{L}_z)]u^2 \pm 2\bar{r}_{\text{KdS}} u + 1}{\sqrt{c_5} \; \Delta_u \sqrt{R_u}} |u^3| du, \quad (4.3.48)$$

where R_u and c_5 are defined in Sec. 4.3.2, Eq. (4.3.37), and $\frac{1}{u^4} \Delta_u = \Delta_{\bar{r} = \pm \frac{1}{u} + \bar{r}_{\text{KdS}}}$, i.e.

$$\Delta_u = (u^2(1 - \bar{\Lambda}\bar{r}_{\text{KdS}}^2) \mp 2\bar{r}_{\text{KdS}} \bar{\Lambda} u - \bar{\Lambda})(u^2(\bar{r}_{\text{KdS}}^2 + \bar{a}^2) \pm 2\bar{r}_{\text{KdS}} u + 1) \mp 2u^3 - 2\bar{r}_{\text{KdS}} u^4, \quad (4.3.49)$$

which is a polynomial of degree 4 in u. Note that for geodesic motion, the coordinate $\bar{r}$ is always contained in an interval bounded by two adjacent real zeros of the polynomial R or by a real zero and infinity. This implies that $u = \pm(\bar{r} - \bar{r}_{\text{KdS}})^{-1}$ for a real zero $\bar{r}_{\text{KdS}}$ of $\bar{R}_{\text{KdS}}$ does not change sign on the integration path and, therefore, we can neglect the absolute value of u appearing in the integrand if we multiply the hole integral with $\text{sign}(u_0) = \frac{u_0}{|u_0|}$. Consequently

$$\frac{I_r}{\bar{a}} = \mp \frac{|u_0|}{u_0} \int_{u_0}^{u} \frac{[\bar{r}_{\text{KdS}}^2 + \bar{a}(\bar{a}E - \bar{L}_z)]u^2 \pm 2\bar{r}_{\text{KdS}} u + 1}{\sqrt{c_5} \; \Delta_u \sqrt{R_u}} u^3 du. \quad (4.3.50)$$

The second step is a partial fraction decomposition of the integrand (neglecting $\sqrt{R_u}^{-1}$) which simplifies I_r to

$$\mp \frac{\sqrt{c_5}|u_0|}{\bar{a} u_0} I_r = C_1 \int_{u_0}^{u} \frac{u \, du}{\sqrt{R_u}} + C_0 \int_{u_0}^{u} \frac{du}{\sqrt{R_u}} + \sum_{i=1}^{4} C_{2,i} \int_{u_0}^{u} \frac{du}{(u-u_i)\sqrt{R_u}}, \quad (4.3.51)$$

4. Geodesics in axially symmetric space-times

where u_i, $1 \leq i \leq 4$ denote the zeros of Δ_u and $C_0, C_1, C_{2,i}$ are the coefficients of the partial fractions, which may be calculated by a computer algebra system and depend on the parameters as well as the zero $\bar{r}_{\text{KdS}}$ of $\bar{R}_{\text{KdS}}$.

The first two integrals in (4.3.51) are of first kind and can be expressed in terms of γ analogous to Sec. 4.3.2, Eq. (4.3.36), i.e.

$$\int_{u_0}^{u} \frac{u \, du}{\sqrt{R_u}} = \sqrt{c_5}(\gamma - \gamma_0), \tag{4.3.52}$$

$$\int_{u_0}^{u} \frac{du}{\sqrt{R_u}} = \int_{u_0}^{\infty} \frac{du}{\sqrt{R_u}} + \int_{\infty}^{u} \frac{du}{\sqrt{R_u}}$$
$$= -f(\sqrt{c_5}\gamma_0 - \gamma_{\bar{r},\text{in}}) + f(\sqrt{c_5}\gamma - \gamma_{\bar{r},\text{in}}), \tag{4.3.53}$$

where again $\gamma_{\bar{r},\text{in}} = \sqrt{c_5}\gamma_0 + \int_{u_0}^{\infty} \frac{u \, du}{\sqrt{R_u}}$ with $u_0 = \pm(\bar{r}_0 - \bar{r}_{\text{KdS}})^{-1}$ only depends on the initial values γ_0 and u_0, and f describes the θ-divisor, i.e. $\sigma\left((f(z), z)^t\right) = 0$.

The four integrals in (4.3.51) containing $(u - u_i)^{-1}$ are in general of third kind and can be expressed in terms of the canonical integral of third kind $\int dP(x_1, x_2)$ defined in (2.2.10). In particular, we get

$$\int_{u_0}^{u} \frac{du}{(u - u_i)\sqrt{R_u}} = \frac{1}{+\sqrt{R_{u_i}}} \int_{u_0}^{u} dP(u_i^+, u_i^-), \tag{4.3.54}$$

where $u_i^+ = (u_i, \sqrt{R_{u_i}})$ is the pole u_i located on the positive branch of the square root and $u_i^- = (u_i, -\sqrt{R_{u_i}})$ is the pole u_i located on the negative branch of the square root. With Eqs. (2.4.12) and (2.4.15) it can be inferred that

$$\int_{u_0}^{u} dP(u_i^+, u_i^-) = \frac{1}{2}\log \frac{\sigma(\int_{\infty}^{u} d\vec{z} - 2\int_{\infty}^{u_i^+} d\vec{z})}{\sigma(\int_{\infty}^{u} d\vec{z} - 2\int_{\infty}^{u_i^-} d\vec{z})} - \frac{1}{2}\log \frac{\sigma(\int_{\infty}^{u_0} d\vec{z} - 2\int_{\infty}^{u_i^+} d\vec{z})}{\sigma(\int_{\infty}^{u_0} d\vec{z} - 2\int_{\infty}^{u_i^-} d\vec{z})}$$
$$- \left(\int_{u_0}^{u} d\vec{z}\right)^t \left(\int_{u_i^-}^{u_i^+} d\vec{r}\right), \tag{4.3.55}$$

where $d\vec{z}$, $d\vec{r}$ are the vectors of the canonical differentials of the first and second kind defined in (2.2.2) and (2.2.3). Finally, we rewrite (4.3.55) in terms of the affine parameter γ. By (4.3.52) and (4.3.53) we can express $\int_{u_0}^{u} d\vec{z}$ as well as the arguments of the σ functions $\int_{\infty}^{u} d\vec{z} = \int_{u_0}^{u} d\vec{z} - \int_{\infty}^{u_0} d\vec{z}$ as functions of γ. If we define $w = w(\gamma) = \sqrt{c_5}\gamma - \gamma_{\bar{r},\text{in}}$ and $w_0 = w(\gamma_0)$ the integral I_r is given by

$$I_r = \mp \frac{\bar{a}u_0}{\sqrt{c_5}|u_0|} \Bigg\{ C_1(w - w_0) + C_0(f(w) - f(w_0))$$
$$+ \sum_{i=1}^{4} \frac{C_{2,i}}{\sqrt{R_{u_i}}} \left[\frac{1}{2}\log \frac{\sigma(W^+(w))}{\sigma(W^-(w))} - \frac{1}{2}\log \frac{\sigma(W^+(w_0))}{\sigma(W^-(w_0))} \right.$$
$$\left. - \left(f(w) - f(w_0), w - w_0\right) \left(\int_{u_i^-}^{u_i^+} d\vec{r}\right) \right] \Bigg\}, \tag{4.3.56}$$

112

4.3. Kerr-de Sitter space-time

where $W^\pm(w) := (f(w), w)^t - 2\int_\infty^{u_i^\pm} d\bar{z}$ and the sign has to be chosen according to the initial substitution $\bar{r} = \pm\frac{1}{u} + \bar{r}_{\text{KdS}}$.

t motion

The equation for t (4.3.11)

$$\frac{1}{\chi^2}\frac{d\bar{t}}{d\gamma} = \frac{\bar{r}^2 + \bar{a}^2}{\Delta_{\bar{r},\text{KdS}}}\mathbb{P}(r) - \frac{\bar{a}}{\Delta_\theta}\mathbb{T}(\theta) \tag{4.3.11}$$

has the same structure as the equation for the φ motion. An integration yields

$$\begin{aligned}\bar{t} - \bar{t}_0 &= \chi^2\left[\int_{\gamma_0}^\gamma \frac{\bar{r}^2 + \bar{a}^2}{\Delta_{\bar{r},\text{KdS}}}\mathbb{P}(r)d\gamma - \int_{\gamma_0}^\gamma \frac{\bar{a}}{\Delta_\theta}\mathbb{T}(\theta)d\gamma\right] \\ &= \chi^2\left[\int_{\bar{r}_0}^{\bar{r}} \frac{(\bar{r}^2 + \bar{a}^2)\mathbb{P}(r)}{\Delta_{\bar{r},\text{KdS}}\sqrt{\tilde{R}_{\text{KdS}}(\bar{r})}}d\bar{r} - \bar{a}\int_{\theta_0}^\theta \frac{\mathbb{T}(\theta)}{\Delta_\theta\sqrt{\tilde{\Theta}_{\text{KdS}}(\theta)}}d\theta\right] \\ &= \chi^2\left[\tilde{I}_r - \bar{a}\tilde{I}_\theta\right].\end{aligned} \tag{4.3.57}$$

Because we already demonstrated the solution procedure, we only give here the results for the most general case.

If $4\nu\Theta_\nu$, where Θ_ν is defined in (4.3.15), has only simple zeros the solution of the θ dependent part is given by

$$\frac{a_3}{|a_3|}\tilde{I}_\theta = (\bar{a}E - \bar{L}_z)(v - v_0)$$

$$-\sum_{i=1}^2 \frac{\bar{a}a_3(\chi E - \bar{a}\bar{\Lambda}\bar{L}_z)}{4\wp'(v_i)}\left[\zeta(v_i)(v - v_0) + \log\sigma(v - v_i) - \log\sigma(v_0 - v_i)\right] \tag{4.3.58}$$

where $\wp(v_1) = \frac{a_2}{12} - \frac{1}{4}\bar{a}^2\bar{\Lambda}a_3 = \wp(v_2)$, a_3, a_2 are defined in (4.3.28), and $v = v(\gamma) = \gamma - \gamma_{\theta,\text{in}}$, $v_0 = v(\gamma_0)$ as in (4.3.43).

If we consider light, i.e. $\epsilon = 0$, the solution for the $\bar{r}$ dependent part is given by

$$\frac{b_3}{|b_3|}\tilde{I}_r = \sum_{i=1}^4\sum_{j=1}^2 \frac{\tilde{C}_i}{\wp'(v_{ij})}\left[\zeta(v_{ij})(v - v_0) + \log\sigma(v - v_{ij}) - \log\sigma(v_0 - v_{ij})\right]$$

$$- \frac{(\bar{r}_{\text{KdS}}^2 + \bar{a}^2)((\bar{r}_{\text{KdS}}^2 + \bar{a}^2)E - \bar{a}\bar{L}_z)}{\Delta_{\bar{r}=\bar{r}_{\text{KdS}}}'}(v - v_0), \tag{4.3.59}$$

where the notation is as in (4.3.34), $\tilde{C}_i$ are the coefficients of the partial fractions $(y - y_i)^{-1}$ with the four zeros y_i of $\Delta_{y(\bar{r})}$, and $\wp(v_{i1}) = y_i = \wp(v_{i2})$.

4. Geodesics in axially symmetric space-times

If $\bar{R}_{\text{KdS}}$ has only simple zeros and we consider timelike geodesics $\epsilon = 1$ the solution of the $\bar{r}$ dependent part is given by

$$\tilde{I}_r = \frac{u_0}{|u_0|\sqrt{c_5}} \left\{ \tilde{C}_1(w - w_0) + \tilde{C}_0(f(w) - f(w_0)) \right.$$

$$+ \sum_{i=1}^{4} \frac{\tilde{C}_{2,i}}{\sqrt{R_{u_i}}} \left[\frac{1}{2} \log \frac{\sigma(W^+(w))}{\sigma(W^-(w))} - \frac{1}{2} \log \frac{\sigma(W^+(w_0))}{\sigma(W^-(w_0))} \right.$$

$$\left. \left. - (f(w) - f(w_0), w - w_0) \left(\int_{u_i^-}^{u_i^+} d\bar{r} \right) \right] \right\} \quad (4.3.60)$$

where the notation is as in (4.3.56) and $\tilde{C}_0, \tilde{C}_1, \tilde{C}_{2,i}$ are the coefficients of the partial fractions $(u - u_i)^{-1}$.

4.3.3 Analytic expressions for observables

For the understanding of characteristic features of a space-time by measurements of geodesics it is crucial to identify certain theoretical quantities of observables. For flyby orbits, this can be the deflection angle of the geodesic whereas for bound orbits it is of interest to determine the orbital frequencies as well as the periastron shift (see also [79]) and the Lense-Thirring effect. The latter can be defined as the motion of the nodes where the orbit of a test particle or light intercepts the equatorial plane. This motion is caused by the g_{0i} components of the space-time metric. In the weak field regime the Lense-Thirring effect becomes visible by a precession of the orbital plane, see Fig. 4.19 for an obvious example. This orbital precession has been confirmed within an accuracy of about 10% by the LAGEOS (Laser Geodynamics Satellite) mission [80] [1].

Let us first consider flyby orbits. The deflection angle of such an orbit depends on the two values $\gamma_\infty^\pm$ of the Mino time for which $\bar{r}(\gamma_\infty^\pm) = \infty$. These are given by

$$\gamma_\infty^\pm = \int_{\bar{r}_0}^{\infty} \frac{d\bar{r}}{\sqrt{\bar{R}_{\text{KdS}}}} = \frac{1}{\sqrt{c_5}} \left(\int_{u_0}^{0} \frac{u du}{\sqrt{R_u}} - \gamma_0 \right) \quad (4.3.61)$$

for the two branches of $\sqrt{R_u}$. Therefore, we can calculate the values of θ and φ for $\bar{r} \to \infty$ that are taken by this flyby orbit by $\theta^\pm = \theta(\gamma_\infty^\pm)$ and $\varphi^\pm = \varphi(\gamma_\infty^\pm)$. The deflection angles are then given by $\Delta\theta = \theta^+ - \theta^-$ and $\Delta\varphi = \varphi^+ - \varphi^-$.

---

[1] Another method to observe the influence of the gravitomagnetic components g_{0i} is through the precession of gyroscopes also known as Schiff effect [81]. Such a measurement has been carried through by Gravity Probe B [82]. While the Lense-Thirring effect is an orbital effect involving the motion of the whole trajectory thus constituting a global measurement, the Schiff effect describes the dragging of local inertial frames due to the existence of the g_{0i} components. For more on gravitomagnetic effects see [83, 84].

4.3. Kerr-de Sitter space-time

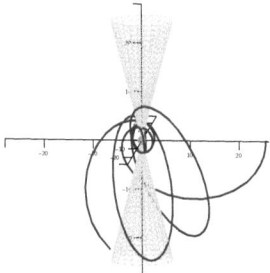

Figure 4.19: Precession of the orbital plane for a bound orbit with $\bar{\Lambda} = 10^{-5}$, $\bar{a} = 0.8$, $\bar{K} = 18$, $\bar{L}_z = -1$, and $E^2 = 0.96$ (region (IIIb)). As in Fig. 4.15, the cones and spheres correspond to extremal θ and horizons.

For bound orbits we can identify three orbital frequencies Ω_r, Ω_θ and Ω_φ associated with the coordinates r, θ and φ. The precessions of the orbital ellipse and the orbital plane, which in the weak field regime can be identified with the periastron advance and the Lense-Thirring effect, respectively, are induced by mismatches of these orbital frequencies. More precisely, the orbital ellipse precesses at $\Omega_\varphi - \Omega_r$ and the orbital plane at $\Omega_\varphi - \Omega_\theta$.

Let us consider the orbital frequency Ω_r. For bound orbits the coordinate $\bar{r}$ is contained in an interval $\bar{r}_p \leq \bar{r} \leq \bar{r}_a$ with the peri- and apoapsis distances $\bar{r}_p$ and $\bar{r}_a$. The orbital period $\omega_{\bar{r}}$ defined by $\bar{r}(\gamma + \omega_{\bar{r}}) = \bar{r}(\gamma)$ is then given by a complete revolution from $\bar{r}_p$ to $\bar{r}_a$ and back (with reversed sign of the square root) to $\bar{r}_p$,

$$\omega_{\bar{r}} = 2 \int_{\bar{r}_p}^{\bar{r}_a} \frac{d\bar{r}}{\sqrt{\bar{R}_{\text{KdS}}}}. \tag{4.3.62}$$

The orbital frequency of the r motion with respect to γ is then given by $\frac{2\pi}{\omega_{\bar{r}}}$. For the calculation of Ω_r, which represents the orbital frequency with respect to t, we need in addition the average rate Γ at which t accumulates with γ. This will be determined below.

For the calculation of the orbital frequency Ω_θ we have to determine the orbital period ω_θ such that $\theta(\gamma + \omega_\theta) = \theta(\gamma)$. The θ motion is likewise bounded by $\theta_{\min} \leq \theta \leq \theta_{\max}$ for two real zeros $\theta_{\min}, \theta_{\max} \in [0, \pi]$ of $\bar{\Theta}_{\text{KdS}}$ and, therefore,

$$\omega_\theta = 2 \int_{\theta_{\min}}^{\theta_{\max}} \frac{d\theta}{\sqrt{\bar{\Theta}_{\text{KdS}}}}. \tag{4.3.63}$$

Again, the orbital frequency of the θ motion with respect to γ is given by $\frac{2\pi}{\omega_\theta}$.

4. Geodesics in axially symmetric space-times

The orbital periods of the remaining coordinates t and φ has to be treated somewhat differently because they depend on both $\bar{r}$ and θ. The solutions $t(\gamma)$ and $\varphi(\gamma)$ consist of two different parts, one which represents the average rates Γ and Y_φ at which t and φ accumulate with γ and one which represents oscillations around it with periods $\omega_{\bar{r}}$ and ω_θ [85, 42]. The periods Γ and Y_φ can be calculated by [42]

$$\Gamma = \frac{2}{\omega_{\bar{r}}} \int_{\bar{r}_p}^{\bar{r}_a} \frac{(\bar{r}^2 + \bar{a}^2)\mathbb{P}(r)}{\Delta_{\bar{r},\text{KdS}}\sqrt{\bar{R}_{\text{KdS}}}} d\bar{r} - \frac{2}{\omega_\theta} \int_{\theta_{\min}}^{\theta_{\max}} \frac{\bar{a}\mathbb{T}(\theta)d\theta}{\Delta_\theta \sqrt{\bar{\Theta}_{\text{KdS}}}}, \tag{4.3.64}$$

$$Y_\varphi = \frac{2}{\omega_{\bar{r}}} \int_{\bar{r}_p}^{\bar{r}_a} \frac{\bar{a}\mathbb{P}(r)}{\Delta_{\bar{r},\text{KdS}}\sqrt{\bar{R}_{\text{KdS}}}} dr - \frac{2}{\omega_\theta} \int_{\theta_{\min}}^{\theta_{\max}} \frac{\mathbb{T}(\theta)d\theta}{\Delta_\theta \sin^2\theta \sqrt{\bar{\Theta}_{\text{KdS}}}}. \tag{4.3.65}$$

The orbital frequencies Ω_r, Ω_θ, and Ω_φ are then given by

$$\Omega_r = \frac{2\pi}{\omega_{\bar{r}}} \frac{1}{\Gamma}, \quad \Omega_\theta = \frac{2\pi}{\omega_\theta} \frac{1}{\Gamma}, \quad \Omega_\varphi = \frac{Y_\varphi}{\Gamma}. \tag{4.3.66}$$

If the integral expressions for ω_r and the $\bar{r}$ dependent parts of Γ and Y_φ degenerate to elliptic or elementary type, i.e. if we consider light or $\bar{R}_{\text{KdS}}$ possesses multiple zeros, we can find analytical expressions for (4.3.62), (4.3.64), and (4.3.65) with the techniques presented in [42]. If $\bar{R}_{\text{KdS}}$ has only simple zeros and $\epsilon = 1$, the integral $\omega_{\bar{r}}$ is an entry of the fundamental period matrix 2ω which enters in the definition of the period lattice of the holomorphic differentials $d\vec{z}$, $\{2\omega v + 2\omega'v \mid v, v' \in \mathbb{Z}^2\}$. The more complicated integrals involving $\bar{R}_{\text{KdS}}$ in (4.3.64) and (4.3.65) can be rewritten in terms of periods of the differentials of the second kind $d\vec{r}$ and of the third kind $dP(x_1, x_2)$ by a decomposition in partial fractions. In this way, expressions for ω_r, Γ and Y_φ which are totally analogous to the elliptic case can be obtained.

It follows that the periastron advance is given by

$$\Delta_{\text{peri}} = \Omega_\varphi - \Omega_r = \left(Y_\varphi - \frac{2\pi}{\omega_{\bar{r}}}\right)\frac{1}{\Gamma} \tag{4.3.67}$$

and the Lense-Thirring effect by

$$\Delta_{\text{Lense-Thirring}} = \Omega_\varphi - \Omega_\theta = \left(Y_\varphi - \frac{2\pi}{\omega_\theta}\right)\frac{1}{\Gamma}. \tag{4.3.68}$$

Another way to access information encoded in the orbits is through a frequency decomposition of the whole orbit [85]. This could be the aim of future work.

4.4 Plebański-Demiański space-times

The methods for analysing geodesic motion and analytically solving the equations of motions developed and applied in this book are not limited to the space-times presented so far in this and the foregoing chapter. Indeed, they can also be used for the most general space-times with separable Hamilton-Jacobi equation, which are the electrovac type-D space-times without acceleration of the gravitating object [72, 73, 74]. The Plebański-Demiański black hole solutions exhausts all electrovac type-D space-times and, thus, it is possible to explicitly and analytically solve the geodesic equations in all space-times, where they are integrable, if this can be done in Plebański-Demiański space-time without acceleration. This space-time is described by [86, 87]

$$ds^2 = \frac{\Delta_{r,\text{PD}}}{\rho_{\text{PD}}^2}\left(dt - (a\sin^2\theta + 2n\cos\theta)d\varphi\right)^2 - \frac{\rho_{\text{PD}}^2}{\Delta_{r,\text{PD}}}dr^2$$
$$- \frac{\Delta_{\theta,\text{PD}}}{\rho_{\text{PD}}^2}\sin^2\theta(adt - (r^2+a^2+n^2)d\varphi)^2 - \frac{\rho_{\text{PD}}^2}{\Delta_{\theta,\text{PD}}}d\theta^2 \quad (4.4.1)$$

with

$$\rho_{\text{PD}}^2 = r^2 + (n - a\cos\theta)^2,$$
$$\Delta_{\theta,\text{PD}} = 1 + \tfrac{1}{3}a^2\Lambda\cos^2\theta - \tfrac{4}{3}\Lambda an\cos\theta \quad (4.4.2)$$
$$\Delta_{r,\text{PD}} = \left(1 - \frac{\Lambda}{3}r^2 - \Lambda n^2\right)(r^2+a^2-n^2) - 2Mr + Q_e^2 + Q_m^2 - \frac{4}{3}\Lambda n^2 r^2$$

where M is the mass, $a = J/M$ the angular momentum per mass, Λ the cosmological constant, n is the NUT charge, Q_e is the electric, and Q_m the magnetic charge of a gravitating source. The axially symmetric space-times considered so far in this chapter emerge from (4.4.1) in the Boyer-Lindquist coordinates $t \to t\chi^{-1}$, $\varphi \to \varphi\chi^{-1}$, where $\chi = 1 + \tfrac{1}{3}a^2\Lambda$, and by setting $n = 0$, $Q_e = 0$, $Q_m = 0$, and in the case of Kerr space-time also $\Lambda = 0$. In Plebański-Demiański space-times there are four constants of motion, namely $\epsilon = g_{\mu\nu}\dot{x}^\mu\dot{x}^\nu$ with $\epsilon = 1$ for timelike and $\epsilon = 0$ for null geodesics, conserved energy per unit mass E given by the generalized momenta p_t, conserved angular momentum per unit mass in z direction L_z given by p_φ, and the Carter constant K obtained by the separation process of the Hamilton-Jacobi equation. It is convenient to introduce the dimensionless quantities

$$\bar{r} = \frac{r}{M}, \quad \bar{t} = \frac{t}{M}, \quad \bar{a} = \frac{a}{M}, \quad \bar{L}_z = \frac{L_z}{M}, \quad \bar{K} = \frac{K}{M^2},$$
$$\bar{\Lambda} = \frac{1}{3}\Lambda M^2, \quad \bar{n} = \frac{n}{M}, \quad \bar{Q}_e = \frac{Q_e}{M}, \quad \bar{Q}_m = \frac{Q_m}{M} \quad (4.4.3)$$

and accordingly

$$\Delta_{\bar{r},\text{PD}} = \left(1 - \bar{\Lambda}(\bar{r}^2 + 3\bar{n}^2)\right)(\bar{r}^2 + \bar{a}^2 - \bar{n}^2) - 2\bar{r} + \bar{Q}_e^2 + \bar{Q}_m^2 - 4\bar{\Lambda}\bar{n}^2\bar{r}^2,$$
$$(\Delta_{r,\text{PD}} = M^2\Delta_{\bar{r},\text{PD}}), \quad (4.4.4)$$
$$\Delta_{\theta,\text{PD}} = 1 + \bar{a}^2\bar{\Lambda}\cos^2\theta - 4\bar{a}\bar{n}\bar{\Lambda}\cos\theta,$$

with which the geodesic equation can be reduced to ordinary differential equations. (Here we not consider the motion of charged test particles which can be treated analogously, see [57].) In addition,

4. Geodesics in axially symmetric space-times

these equations can be decoupled in terms of the normalized Mino time $\gamma = \lambda M$, where λ is related to the proper time τ by $\frac{d\tau}{d\lambda} = \rho_{\text{PD}}^2$, yielding

$$\left(\frac{d\bar{r}}{d\gamma}\right)^2 = \mathbb{P}^2(\bar{r}) - \Delta_{\bar{r},\text{PD}}(\epsilon \bar{r}^2 + \bar{K}) =: \bar{R}_{\text{PD}}(\bar{r}), \tag{4.4.5}$$

$$\left(\frac{d\xi}{d\gamma}\right)^2 = \Delta_{\theta,\text{PD}}(1-\xi^2)\left(\bar{K} - \epsilon(\bar{n} - \bar{a}\xi)^2\right) - \mathbb{T}(\theta)_{\text{PD}}^2 =: \Theta_\xi(\xi), \tag{4.4.6}$$

$$\frac{d\varphi}{d\gamma} = \frac{\bar{a}}{\Delta_{\bar{r},\text{PD}}}\mathbb{P}(r)_{\text{PD}} - \frac{1}{\Delta_{\theta,\text{PD}}\sin^2\theta}\mathbb{T}(\theta)_{\text{PD}}, \tag{4.4.7}$$

$$\frac{d\bar{t}}{d\gamma} = \frac{\bar{r}^2 + \bar{a}^2 + \bar{n}^2}{\Delta_{\bar{r},\text{PD}}}\mathbb{P}(r)_{\text{PD}} - \frac{\bar{a}\sin^2\theta + 2\bar{n}\cos\theta}{\Delta_{\theta,\text{PD}}\sin^2\theta}\mathbb{T}(\theta)_{\text{PD}}, \tag{4.4.8}$$

where $\xi = \cos\theta$ and

$$\begin{aligned}\mathbb{P}(\bar{r})_{\text{PD}} &= (\bar{r}^2 + \bar{a}^2 + \bar{n}^2)E - \bar{a}\bar{L}_z, \\ \mathbb{T}(\theta)_{\text{PD}} &= (\bar{a}\sin^2\theta + 2\bar{n}\cos\theta)E - \bar{L}_z,\end{aligned} \tag{4.4.9}$$

which reduce to (4.2.12) in the case of Kerr or Kerr-de Sitter space-time. Both of the polynomials $\bar{R}$ and Θ_ξ are in general of degree 6 and, thus, Eqs. (4.4.5) and (4.4.6) are of hyperelliptic type and first kind. After inserting $\bar{r}(\gamma)$ and $\theta(\gamma)$ into Eqs. (4.4.7) and (4.4.8) they will be of hyperelliptic type and third kind. Therefore, the equations of motions in Plebański-Demiański space-times can be completely analytically solved with the methods used earlier in this chapter (as also shown in [77]).

A standard substitution $\xi = \pm\frac{1}{u} + \xi_\Theta$, where ξ_Θ is a zero of Θ_ξ, transforms Eq. (4.4.6) to the standard form $(u\frac{du}{d\gamma})^2 = c_\theta \Theta_u$ for a constant c_θ and a polynomial Θ_u of degree 5. The solution of Eq. (4.4.6) for θ is then given by

$$\theta(\gamma) = \arccos\left(\mp\frac{\sigma_2}{\sigma_1}\left(\frac{f(\gamma - \gamma_{\theta,\text{in}})}{\gamma - \gamma_{\theta,\text{in}}}\right) + \xi_\Theta\right), \tag{4.4.10}$$

where the sign in (4.4.10) is the same as in the substitution $\xi = \pm\frac{1}{u} + \xi_\Theta$. Note that in contrast to the axially symmetric space-times considered so far in this chapter, here the motion in θ direction is in general not symmetric with respect to the equatorial plane.

The same type of substitution $\bar{r} = \pm\frac{1}{u} + \bar{r}_R$ with a zero $\bar{r}_R$ of $\bar{R}_{\text{PD}}$ casts $\bar{R}_{\text{PD}}$ in the standard form $(u\frac{du}{d\gamma})^2 = c_{\bar{r}} R_u$, where $c_{\bar{r}}$ is a constant and R_u a polynomial of degree 5. The solution of (4.4.5) can be written as

$$\bar{r}(\gamma) = \mp\frac{\sigma_2}{\sigma_1}\left(\frac{f(\gamma - \gamma_{\bar{r},\text{in}})}{\gamma - \gamma_{\bar{r},\text{in}}}\right) + \bar{r}_R, \tag{4.4.11}$$

where the sign again depends on the one chosen in the substitution. Here and in (4.4.10) the function f describes the theta divisor, i.e. $\sigma((\frac{f(x)}{x});\tau) = 0$, which depends on the normalized period matrix τ which in turn depends on the polynomials Θ_ξ for the θ motion and on $\bar{R}_{\text{PD}}$ for the $\bar{r}$ motion.

4.4. Plebański-Demiański space-times

The equations for φ and $\bar{t}$ can be solved analogously to (4.3.40) and (4.3.57). The solution for (4.4.7) is given by

$$\varphi - \varphi_0 = I_{\varphi,\bar{r}} + I_{\varphi,\theta} \,. \tag{4.4.12}$$

Here the integral $I_{\varphi,\bar{r}}$ can be solved by

$$I_{\varphi,\bar{r}} = -\frac{\bar{a}u_0}{\sqrt{c_{\bar{r}}}|u_0|}\bigg\{ C_1^{\bar{r}}(w - w_0) + C_0^{\bar{r}}(f(w) - f(w_0))$$
$$+ \sum_{i=1}^{4} \frac{C_{2,i}^{\bar{r}}}{\sqrt{R_u(u_i)}} \bigg[\frac{1}{2}\log\frac{\sigma(W^+(w))}{\sigma(W^-(w))} - \frac{1}{2}\log\frac{\sigma(W^+(w_0))}{\sigma(W^-(w_0))}$$
$$- (f(w) - f(w_0), w - w_0) \bigg(\int_{u_i^-}^{u_i^+} d\vec{r}\bigg)\bigg]\bigg\} \,, \tag{4.4.13}$$

where the notation is as in (4.3.56), and the integral $I_{\varphi,\theta}$ by

$$I_{\varphi,\theta} = -\frac{u_0}{\sqrt{c_{\theta}}|u_0|}\bigg\{ C_1^{\theta}(w - w_0) + C_0^{\theta}(f(w) - f(w_0))$$
$$+ \sum_{i=1}^{4} \frac{C_{2,i}^{\theta}}{\sqrt{\Theta_u(u_i)}} \bigg[\frac{1}{2}\log\frac{\sigma(W^+(w))}{\sigma(W^-(w))} - \frac{1}{2}\log\frac{\sigma(W^+(w_0))}{\sigma(W^-(w_0))}$$
$$- (f(w) - f(w_0), w - w_0) \bigg(\int_{u_i^-}^{u_i^+} d\vec{r}\bigg)\bigg]\bigg\} \,, \tag{4.4.14}$$

where u_i are the zeros of $\Delta_{\theta=\arccos(\pm u^{-1}-\xi_\Theta),\mathrm{PD}}$, u_0 is the initial value $u_0 = (\cos(\theta_0) - \xi_\Theta)^{-1})$, C_0^{θ}, C_1^{θ}, and $C_{2,i}^{\theta}$ are the coefficients of the partial fractions of u^0, u^1 and $(u-u_i)^{-1}$, respectively, which can be obtained by a computer algebra system. The notation of w, w_0, and $W^{\pm}$ are the same as in (4.3.56). The solution of Eq. (4.4.8) can be written in the same way with the appropriate partial fractions.

Therefore we succeeded in obtaining the complete analytic solution of the geodesic equation in all Plebański-Demiański black hole space-times without acceleration. An analysis of possible orbit types in these space-times is very voluminous due to the large number of parameters, which are the four constants of motion ϵ, E, $\bar{L}_z$, and $\bar{K}$ as well as the five parameters characterizing this class of space-times $\bar{a}$, $\bar{\Lambda}$, $\bar{Q}_e$, $\bar{Q}_m$, and $\bar{n}$ (the mass is absorbed through the rescaling (4.4.3)). However, an analysis for the special cases of spherical symmetric space-times can be found in Chap. 3, for the cases of Kerr and Kerr-de Sitter space-times in this chapter, and for Taub-NUT-de Sitter space-time corresponding to $\bar{a} = 0$, $\bar{Q}_e = 0$, and $\bar{Q}_m = 0$ in [77]. As the Plebański-Demiański black hole solutions exhaust all electrovac type-D solutions and the analytical solutions of the geodesic equations in Plebański-Demiański space-times without acceleration was given above, it can be concluded that the analytic solutions of the geodesic equations in all electrovac type-D space-times without acceleration can explicitly be given.

CHAPTER 5

Summary and Outlook

5.1 Summary

The aim of this book was to study the influence of the cosmological constant on geodesics in the black hole space-times of Schwarzschild, Reissner-Nordström, and Kerr. All calculations were carried out by using analytical methods, which enabled a systematic study of effects and assured an, in principle, unlimited calculational accuracy. Both timelike and null geodesics were discussed including some particular interesting orbits like the innermost stable circular orbit, and analytic expressions for observables were derived.

The foundations of the mathematical methods used in this book are known since the 19th century and were outlined here in the second chapter. In particular, the theory of elliptic function was used already in 1931 to solve geodesic equations in Schwarzschild space-time. However, solutions of the more complicated geodesic equations containing polynomials of a degree up to six, namely those in space-times with a nonvanishing cosmological constant, require the theory of hyperelliptic functions. Here a new aspect enters with the introduction of the theta divisor, which is defined as the set of zeros of a theta function. Usually, it was assumed that the given problem is not defined on the theta divisor, but in the cases considered here the contrary holds and is the key element for the solution method. This application of the theta divisor was first used in [51] for the case of a double pendulum.

In the following chapters, the mathematical methods were applied to the geodesic equation corresponding to different solutions of Einstein's field equations. At first, the geodesic equations in

5.1. Summary

spherically symmetric space-times, namely the Schwarzschild, Reissner-Nordström, and their generalizations to a nonvanishing cosmological constant were considered. Although the geodesic motion in Schwarzschild space-time was analyzed in detail long before by Hagihara [6], a treatise of the same completeness was given only recently for Reissner-Nordström space-times by Slezáková [15], where the analytical solutions of the geodesic equation were given in terms of Jacobi elliptic functions. In this book, not only the almost complete set of orbits was analyzed but also the analytical solutions for the most general cases of geodesic motion for both timelike and null geodesics were derived in terms of Weierstrass elliptic functions. The only types of orbits not considered here are the motions of charged particles treated, for example, in [14].

On the contrary, the complete analytical treatment of geodesic motion in the corresponding space-times with a nonvanishing cosmological constant is entirely new. The complete set of orbits in Schwarzschild-(anti-) de Sitter and Reissner-Nordström-(anti-)de Sitter space-times were classified and the general analytical solution for the geodesic equation presented. In addition, an approximation of the periastron advance of bound orbits for a small cosmological constant was derived in Schwarzschild-de Sitter space-time, applied to the orbital data of Mercury and Quasar QJ287, and in the first case compared to earlier results [40]. As a further application of the analytical results the influence of the cosmological constant on the Pioneer 10 and 11 spacecraft was computed and found to be too small to produce a measureable effect. Finally, it was demonstrated that the methods developed and applied in this chapter can also be used to integrate the geodesic equation in up to seven-dimensional Schwarzschild space-times. This can even be extended to more general higher-dimensional space-times as shown in [53].

In the fourth chapter the methods presented and developed in the foregoing chapters were applied to more general axially symmetric space-times, i.e. to Kerr as well as Kerr-(anti-)de Sitter space-time and even to the wide class of Plebański-Demiański space-times. A major difficulty compared to the spherical symmetric space-times considered before is the need for a fourth constant of motion, the Carter constant [11]. The geodesic motion in Kerr space-time was investigated since the metric was discovered in 1963, but due to the coupled character of the equations and the complexity of the space-time a number of special classes of geodesics were analyzed instead of considering the most general case. Again, a complete analytical treatment was presented only recently by Slezáková [15] using Jacobi elliptic functions. However, Slezáková was obviously not aware of a major simplification of the equations of motion in Kerr space-time first presented in 2003 by Mino [30], which decouples the equations and allows to explicitly solve the equations of motions in terms of an affine parameter. In this book, this approach was used to formulate the most general analytical solution of the geodesic equation in Kerr space-time in terms of Weierstrass elliptic functions depending on an affine parameter, the Mino time. Also, the complete set of orbits for slowly rotating Kerr black holes was classified in terms of the parameters of the test particle or light ray.

5. Summary and Outlook

The generalization of these results to the case of a nonvanishing cosmological constant could in principle be done in the same way as the generalization in the spherically symmetric case. However, two difficulties arise here. The first is connected to the Carter constant derived in the separation process of the Hamilton-Jacobi equation. For a vanishing cosmological constant there are two forms of this constant one of which is directly related to the geometry of the geodesic [14]. Although for a nonvanishing cosmological constant only the other one of these two forms emerges from the separation process, it was shown in this book that for a positive and, to some extend, for a negative cosmological constant the Carter constant is nevertheless related to the geometry of the orbit. The second difficulty is connected to the fact that in contrast to the spherically symmetric case, the radial and latitudinal coordinate can not be directly expressed in terms of the azimuthal angle, which introduces hyperelliptic integrals of third kind. This issue could be solved and, thus, the analytical solution of the geodesic equation in Kerr-(anti-)de Sitter space-time could be presented here for the first time. Furthermore, the differences between the geodesic motion in Kerr and Kerr-de Sitter space-times were analyzed and a procedure for the determination of the last stable spherical and the innermost circular orbit in the equatorial plane was outlined. The properties of the analytical solution were used to determine analytic expressions for observables in Kerr-de Sitter space-time, namely the deflection angle of flyby orbits and the periastron advance as well as the Lense-Thirring effect of bound orbits. Last but not least, the methods used so far were also applied to the general class of Plebański-Demiański electrovac space-times without acceleration of the gravitating object, which exhaust all space-times with integrable equation of motions. Therefore, it was shown that in all space-times with integrable equations of motion an analytical solution can indeed be found.

5.2 Outlook

The work accomplished in this book can be continued in a number of ways, namely by improvements of the used methods, enhanced discussion of geodesics and observables in space-times with nonvanishing cosmological constant, and application of the methods developed here to geodesic motion in more general space-times and the effective one-body formalism.

Improvement of methods A key element of the analytical solution of the geodesic equation in space-times with a nonvanishing cosmological constant is the concept of the theta divisor. However, its definition is rather implicit and not well suited for computations. It would be a major improvement of the solution method to formulate this complex one-dimensional submanifold in terms of charts or as the graph of a differentiable function (using the implicit function theorem). If it will turn out that this differentiable function can not be expressed in a closed form, at least it should be possible to find a series expansion. More precisely, as $\exp(\text{Re}((m+g)^t(i\tau)(m+g)^t))$ becomes small for large $|m|$

(see (2.3.1)), the expression (2.3.4) can be approximated by considering only small $|m|$ and may then be solved for z_1 depending on z_2 or vice versa.

Another important aspect of this method is the definition and computation of the periodicity of the resulting solutions, which is directly connected to the period matrix of the vector of holomorphic differentials on the corresponding Riemann surface. As explained in this book, this periodicity also defines the observables connected with the trajectory of a geodesic, i.e. the periastron advance and the Lense-Thirring effect. Therefore, it is desireable to solve the integrals defining the period matrix in terms of analytic functions. Hints for a realization can be found in [88].

So far the presented analytical solution methods are limited to hyperelliptic problems with an underlying polynomial of degree 6 or lower. However, the equations of motion in some higher-dimensional spherically symmetric space-times and maybe also in some of the generalized space-times discussed below contain polynomials of a higher degree. These types of equations can not be solved using the restriction to the theta divisor discussed in Sec. 3.3.1.2, as this is a manifold of complex dimension two or higher in these cases. Hints for a generalization of the solution method to these cases can be found in [51].

Enhanced discussion of geodesics and observables The results of this book can be viewed as the starting point for the analysis of several features of geodesics in space-times with a nonvanishing cosmological constant not treated here. Although mathematical analogous to the case of slow Kerr-de Sitter a complete discussion of orbits in fast and extreme Kerr-de Sitter space-times may lead to special features. This was carried out for the fast case [89] but should be extended also to the extreme case. It would also be interesting to study bound geodesics crossing $\bar{r} = 0$ (and the Cauchy horizon for positive $\bar{r}$) in general and, in particular, their causal structure. In this context the analysis of closed timelike trajectories is also of interest. In addition, geodesics lying entirely on the axis $\theta = 0, \pi$ or even crossing it were not consider yet. The Boyer-Lindquist form of the Kerr-de Sitter metric used here is not a good choice for considering geodesics which fall through a horizon. Therefore, for future work it would be interesting to use a coordinate-singularity free version of the metric.

Analytic solutions are the starting point for approximation methods for the description of real stellar, planetary, comet, asteroid, or satellite trajectories (see e.g. [90]). In particular, it is possible to derive post-Kerr, post-Schwarzschild, or post-Newton series expansions of analytical solutions. In addition to the series expansion of the periastron advance in Schwarzschild-de Sitter space-time in terms of Λ presented in Sec. 3.3 it would be interesting to derive post-Kerr, post-Schwarzschild, or post-Newton expressions for this and other observables. Due to the high precision, the analytical expressions for observables in space-times with a nonvanishing Λ may be used for comparisons with observations where the influence of the cosmological constant might play a role. This could be the

5. Summary and Outlook

case for stars moving around the galactic center black hole or binary systems with extreme mass ratios where one body serves as test particle.

Due to the, in principle, arbitrary high accuracy of analytic solutions of the geodesic equation they can also serve as test beds for numerical codes for the dynamics of binary systems in the extreme stellar mass ratio case (extreme mass ratio inspirals, EMRIs) and also for the calculation of corresponding gravitational wave templates. For the case of Kerr space-time with a vanishing cosmological constant it has already been shown that gravitational waves from EMRIs can be computed more accurately by using analytical solutions than by numerical integration [42].

Generalized space-times and effective one-body formalism The presented methods for obtaining analytical solutions of geodesic equations can also be applied to other space-times. Indeed, they have already been used to solve the geodesic equations in higher-dimensional static spherically symmetric space-times [53] and for the case of a Schwarzschild black hole pierced by a cosmic string [91]. It will also be interesting to apply the presented methods to higher-dimensional stationary axially symmetric space-times like the Myers-Perry solutions and to generalize the results of [91] to stationary space-times or to a nonvanishing cosmological constant. In particular, the Plebański-Demiański space-time without acceleration treated in Sec. 4.4 and all its special cases could be elaborated [77, 92]. For the case of Taub-NUT and Taub-NUT-de Sitter space-time this work is already in progress [93, 94] including an extensive discussion of geodesic incompleteness.

The same structure of equations is also present in the geodesic equation of the effective one-body formalism of the relativistic two-body problem. The effective metric in this formalism can be described as a perturbed Schwarzschild or Kerr metric, where the perturbation is given in powers of the radial coordinate r [95, 96, 97]. Therefore, it can be expected that the polynomial appearing in the resulting equations of motion will have a higher degree than the corresponding polynomial in the Schwarzschild or Kerr case and, thus, that analytical solutions of these equations will require the use of hyperelliptic instead of elliptic functions. A similar situation can be found in the expressions of axisymmetric gravitational multipole space-times. For example, some types of geodesics in Erez-Rosen space-time, which reduces to the Schwarzschild case if the quadrupole moment is neglected, were already solved analytically assuming a weak field[98, 99]. Probably, the methods presented in this book will be helpful to solve geodesics in space-times with higher order multipoles.

APPENDIX A

Elliptic integrals of third kind

In this appendix the details for the calculation of elliptic integrals of the third kind (see also (2.1.17))

$$\int_{y_1}^{y_2} \frac{f(y)\,dy}{\sqrt{4y^3 - g_2 y - g_3}} = \int_{v_1}^{v_2} f(\wp(v))\,dv \tag{A.0.1}$$

will be given, where g_2, g_3 are the Weierstrass invariants, f is a rational function, and the sign of the square root is chosen according to the sign of $\wp'$. The different choices of f discussed in the following sections correspond to different problems discussed in this book. For solving the integral (A.0.1) the function f will be expressed in terms of the Weierstrass ζ function as well as the Weierstrass $\wp$ function and its derivatives along the lines of Thm. 2.5. The reason is that these functions can be integrated easily since $\zeta' = -\wp$ and $(\log \sigma)' = \zeta$, where σ is the Weierstrass σ-function:

$$\int_\gamma \wp(v - v_0)\,dv = \zeta(\gamma(0) - v_0) - \zeta(\gamma(1) - v_0), \tag{A.0.2}$$

$$\int_\gamma \zeta(v - v_0)\,dv = \log \sigma(\gamma(1) - v_0) - \log \sigma(\gamma(0) - v_0), \tag{A.0.3}$$

where $\gamma(0) = v_1$ and $\gamma(1) = v_2$ (the branches of log will be discussed later).

Although the integration procedure for (A.0.1) is not complicated, often lengthy calculations are necessary to obtain the constants A_i^n defined in Thm. 2.5, i.e. to find the representation of an elliptic function in terms of the Weierstrass ζ function and its derivatives. These calculations are explicitly carried out in this appendix for the case of the post-Schwarzschild periastron advance presented in Sec. 3.3.1.3 and for a generic case which can easily be matched to all other elliptic integrals of third kind discussed in this book.

A. Elliptic integrals of third kind

A.1 General solution procedure

In this section we present the general solution procedure for an elliptic integral of third kind and demonstrate it for an example. We consider

$$\int_{v_0}^{v} \frac{dv}{c_1 \wp(v) - c_2} =: \int_{v_0}^{v} f(v) dv, \tag{A.1.1}$$

where c_1 and c_2 are constant. By using some substitutions and a partial fraction decomposition this generic case can easily be adapted to the integrals (4.2.34), (4.2.40), (4.2.43), the θ dependent part of (4.3.57) and (for null geodesics) (4.3.45) as well as the r dependent part of (4.3.57).

The function f has poles of first order at $v_1, v_2 \in \{2t\omega + 2s\omega' \mid 0 \leq t, s < 1\}$ with the periods $2\omega \in \mathbb{R}, 2\omega' \in \mathbb{C}$ of $\wp$, where $\wp(v_j) = \frac{c_2}{c_1}$. (This means that $v_2 = 2\omega - v_1$ if $v_1 = 2t\omega$, $v_2 = 2\omega' - v_1$ if $v_1 = 2s\omega'$, and $v_2 = 2\omega + 2\omega' - v_1$ else.) Therefore, assuming $v_1 \neq v_2$[1], in a neighborhood of v_j the function f is given by

$$f(v) = \frac{a_j}{v - v_j} + \text{holomorphic part} \tag{A.1.2}$$

whereas $(c_1 \wp(v) - c_2)$ expands in this neighborhood as

$$c_1 \wp(v) - c_2 = c_1 \wp'(v_j)(v - v_j) + \text{higher orders}. \tag{A.1.3}$$

The combination of these two expansions can now be used to determine the constants A_i^n defined in Thm. 2.5. A comparison of coefficients yields

$$1 = f(v)(c_1 \wp(v) - c_2) = a_j c_1 \wp'(v_j) + \text{higher orders} \quad \Rightarrow \quad a_j = \frac{1}{c_1 \wp'(v_j)}. \tag{A.1.4}$$

Because ζ is an elliptic function with a simple pole in 0 with residue 1 this implies that

$$f(v) - \sum_{j=1}^{2} \frac{\zeta(v - v_j)}{c_1 \wp'(v_j)} \tag{A.1.5}$$

is an elliptic function without poles and, therefore (see Thm. 2.2), equal to a constant A, which can be computed using $f(0) = 0$. From this it can be inferred that

$$f(v) = \sum_{j=1}^{2} \frac{\zeta(v - v_j) + \zeta(v_j)}{c_1 \wp'(v_j)}. \tag{A.1.6}$$

By the formula (A.0.3) f can then be integrated to

$$\int_{v_0}^{v} f(v) dv = \sum_{j=1}^{2} \frac{\zeta(v_j)(v - v_0) + \log \sigma(v - v_j) - \log \sigma(v_0 - v_j)}{c_1 \wp'(v_j)}. \tag{A.1.7}$$

---
[1]If $v_1 = v_2$, f has one pole of second order and can be given in terms of $\zeta(v - v_1)$ and additionally $\wp(v - v_1)$.

A.1. General solution procedure

Note that $\wp'(v_j)$ can be expressed in terms of c_1 and c_2 by using the relation

$$\wp'(v_j) = \pm\sqrt{4\wp(v_j)^3 - g_2\wp(v_j) - g_3} = \pm\sqrt{4\frac{c_2^3}{c_1^3} - g_2\frac{c_2}{c_1} - g_3}. \qquad (A.1.8)$$

In general, the elliptic integrals of third kind to be solved in this book arrived at the form (A.1.1) by substitutions of the form $x = g(y)$ and $y = \wp(v)$, where x denotes a space-time coordinate and g a rational function. Furthermore, the substitution $x = g(y)$ was in general applied before to an integral of first kind, with the result that $y = \wp(\gamma - \gamma_{\text{in}})$ for an affine parameter γ (the normalized Mino time) and a constant γ_{in} depending on the initial values of the problem. This allows to substitute $v = \gamma - \gamma_{\text{in}}$ (modulo periods) and accordingly $v_0 = v(\gamma_0)$ in Eq. (A.1.7), what results in

$$\int_{v_0}^{v} f(v)dv = \sum_{j=1}^{2} \frac{1}{c_1 \wp'(v_j)} \left(\zeta(v_j)(\gamma - \gamma_0) + \log\sigma(\gamma - \gamma_{\text{in}} - v_j) - \log\sigma(\gamma_0 - \gamma_{\text{in}} - v_j)\right). \qquad (A.1.9)$$

As an example, consider the integral I_r in Eq. (4.2.34). The corresponding integral of first kind $\int_{\bar{r}_0}^{\bar{r}} \frac{d\bar{r}}{\sqrt{\bar{R}_K}}$ was solved in 4.2.2 with substitutions $\bar{r} = \frac{1}{\xi} + \bar{r}_K$ and $\xi = \frac{1}{a_3}\left(4y - \frac{a_2}{3}\right)$. The solution for y was then given by $y = \wp(\gamma - \gamma_{r,\text{in}})$. As the same substitutions and additionally $y = \wp(v)$ were used to cast the integral I_r in the standard form as a sum of terms (A.1.1), the result $y = \wp(\gamma - \gamma_{r,\text{in}})$ can be used to express I_r in terms of γ.

Example We explicitly demonstrate the generic solution procedure explained above for the conrete integral (4.2.40)

$$I_\theta := -\frac{|a_3|}{a_3}\left[\bar{a}E(v - v_0) + \int_{v_0}^{v} \frac{a_3 \bar{L}_z dv}{4\wp(v) - a_3 - \frac{a_2}{3}}\right],$$

where a_3 and a_2 are some constants.

Here the function f from (A.0.1) is given by

$$f(v) = \frac{1}{4\wp(v) - c}, \qquad (A.1.10)$$

where $c = a_3 + \frac{a_2}{3} = 4\left(\frac{1}{3}(2\bar{a}E(\bar{a}E - \bar{L}_z) - \bar{K} - \epsilon\bar{a}^2) - \bar{a}^2(E^2 - \epsilon)\right)$. The function f has poles of first order at $v_1, v_2 \in \{tw + sw' \mid 0 \leq t, s < 1\}$ with the periods w, w' of $\wp$, where $\wp(v_j) = \frac{c}{4}$. Therefore, in a neighborhood of v_j the function f is given by

$$f(v) = \frac{a_j}{v - v_j} + \text{holomorphic part} \qquad (A.1.11)$$

and the expression $(4\wp(v) - c)$ by

$$4\wp(v) - c = 4\wp'(v_j)(v - v_j) + \text{higher orders}. \qquad (A.1.12)$$

A. Elliptic integrals of third kind

The combination of these two expansions yields

$$1 = f(v)(4\wp(v) - c) = a_j 4\wp'(v_j) + \text{higher orders} \quad \Rightarrow \quad a_j = \frac{1}{4\wp'(v_j)}. \tag{A.1.13}$$

Because ζ is an elliptic function with a simple pole in 0 with residue 1 this implies that

$$f(v) - \sum_{j=1}^{2} \frac{\zeta(v - v_j)}{4\wp'(v_j)} \tag{A.1.14}$$

is an elliptic function without poles, and, therefore, equal to a constant A, which can be computed using $f(0) = 0$. The yields

$$f(v) = \sum_{j=1}^{2} \frac{\zeta(v - v_j) + \zeta(v_j)}{4\wp'(v_j)}. \tag{A.1.15}$$

Then f can be integrated giving

$$\int_{v_0}^{v} f(v) dv = \sum_{j=1}^{2} \frac{1}{\wp'(v_j)} \left(\zeta(v_j)(v - v_0) + \log \sigma(v - v_j) - \log \sigma(v_0 - v_j) \right). \tag{A.1.16}$$

Here $\wp'(v_j)$ can be expressed in terms of c by using $\wp(v_j) = \frac{c}{4}$ and the differential equation (2.1.2). We insert this result in (4.2.40):

$$I_\theta = -\frac{|a_3|}{a_3} \left[\bar{a} E(v - v_0) \right.$$

$$\left. + \frac{a_3 \bar{L}_z}{4} \sum_{j=1}^{2} \frac{1}{\wp'(v_j)} \left(\zeta(v_j)(v - v_0) + \log \sigma(v - v_j) - \log \sigma(v_0 - v_j) \right) \right]. \tag{A.1.17}$$

Because $\wp(v) = y$ and $y(\gamma) = \wp(\gamma - \gamma_{\theta,in})$, which was derived in the solution process of the θ equation (see (4.2.30)), it can be inferred that $v = \gamma - \gamma_{\theta,in}$ modulo periods. This implies

$$I_\theta(\gamma) = -\frac{|a_3|}{a_3} \left[\bar{a} E(\gamma - \gamma_0) \right.$$

$$\left. + \frac{a_3 \bar{L}_z}{4} \sum_{j=1}^{2} \frac{1}{\wp'(v_j)} \left(\zeta(v_j)(\gamma - \gamma_0) + \log \sigma(\gamma - \gamma_j) - \log \sigma(\gamma_0 - \gamma_j) \right) \right] \tag{A.1.18}$$

modulo the periods ω, ω' and the periods of second kind η, η'. (More precisely, these periods are multiplied by the constant factors in this equation. See also (2.1.16)) Here $\wp(\gamma_j - \gamma_{\theta,in}) = \frac{a_3}{4} + \frac{a_2}{12}$.

A.2 Post-Schwarzschild periastron advance

For an integration of the second term on the right hand side in (3.3.40), which can be transformed to (3.3.45), the functions

$$F_1(v) = \frac{1}{P_W(\wp(v))} = \frac{1}{(\wp'(v))^2},$$
$$F_2(v) = \frac{1}{(2\wp(v) + \frac{1}{6})^2 P_W(\wp(v))} = \frac{1}{(2\wp(v) + \frac{1}{6})^2 (\wp'(v))^2} \quad \text{(A.2.1)}$$

have to be integrated along a path γ with $\gamma(0) = v_1$ and $\gamma(1) = v_1 + 2w$, where w is the purely real period of the $\wp$ function. Due to the periodicities

$$\zeta(z + 2w) = \zeta(z) + 2\eta, \quad \text{(A.2.2)}$$
$$\sigma(z + 2w) = e^{2\eta(z+w)+\pi i}\sigma(z), \quad \text{(A.2.3)}$$

Eqs. (A.0.2) and (A.0.3) can be rewritten as

$$\int_\gamma \wp(v - v_0) dv = -2\eta, \quad \text{(A.2.4)}$$
$$\int_\gamma \zeta(v - v_0) dv = 2\eta(v_1 - v_0 + w) + \pi i + 2\pi i k \quad \text{(A.2.5)}$$

for a $k \in \mathbb{Z}$, where η is the period of second kind.

Integration of F_1: The function F_1 only possesses poles of second order in $\rho_1 = w'$, $\rho_2 = w' + w$, $\rho_3 = w$ (see Fig. 3.14). In a neighborhood of ρ_j, the function F_1 can be expanded as

$$F_1(v) = \frac{a_{j2}}{(v - \rho_j)^2} + \frac{a_{j1}}{v - \rho_j} + \text{holomorphic part}. \quad \text{(A.2.6)}$$

Since $\wp'(\rho_j + z)^2 = \wp'(\rho_j - z)^2$ for all j and z, F_1 is symmetric with respect to all ρ_j and, therefore, depends only on even powers of $(v - \rho_j)$ implying $a_{j1} = 0$. The constant a_{j2} can be evaluated by a comparison of coefficients. For this, we note that $\wp'(\rho_j) = 0 = \wp'''(\rho_j)$ and, thus,

$$\wp'(v) = \wp''(\rho_j)(v - \rho_j) + \sum_{i=3}^\infty c_i(v - \rho_j)^i \quad \text{(A.2.7)}$$

in a neighborhood of ρ_j and for some constants c_i. If we square both sides of the equation, we see that $\wp'^2$ contains only $(v - \rho_j)^2$ and powers of $(v - \rho_j)$ larger than 4. It follows

$$1 = F_1(v)\wp'(v)^2 = a_{j2}\wp''(\rho_j)^2 + \text{higher powers of } (v - \rho_j). \quad \text{(A.2.8)}$$

A. Elliptic integrals of third kind

This implies $a_{j2} = \frac{1}{\wp''(\rho_j)^2}$ for all j. The function $\wp(v - \rho_j)$ has only one pole of second order in ρ_j with zero residue. Therefore, the difference

$$F_1(v) - \sum_{j=1}^{3} a_{j2}\wp(v - \rho_j) \tag{A.2.9}$$

is a holomorphic elliptic function and, thus, constant by Thm. 2.2. This yields

$$F_1(v) = \sum_{j=1}^{3} a_{j2}\wp(v - \rho_j) + c_1 . \tag{A.2.10}$$

The constant c_1 can be determined by $F_1(0) = 0$ using the relation $\wp(-\rho_j) = \wp(\rho_j) = y_j$

$$c_1 = -\sum_{j=1}^{3} a_{j2} y_j . \tag{A.2.11}$$

In summary, we obtain

$$\begin{aligned}
\oint_A \frac{dy}{P_W(y)\sqrt{P_W(y)}} &= \int_\gamma F_1(v) dv \\
&= \int_\gamma \sum_{j=1}^{3} a_{j2}(\wp(v - \rho_j) - y_j) dv \\
&= \sum_{j=1}^{3} \frac{1}{\wp''(\rho_j)^2} \left(\int_\gamma \wp(v - \rho_j) dv - 2\omega y_j \right) \\
&= \sum_{j=1}^{3} \frac{1}{\wp''(\rho_j)^2} (-2\eta - 2\omega y_j) .
\end{aligned} \tag{A.2.12}$$

Integration of F_2: The function F_2 possesses poles of second order in $\rho_1 = \omega'$, $\rho_2 = \omega' + \omega$, $\rho_3 = \omega$ and in all $v_0 \in \mathbb{R}$ such that $\wp(v_0) = -\frac{1}{12}$. Since we assumed that the considered orbit is bound, all zeros of P_{SdS} have to be positive and, thus, $-\frac{1}{12} < y_1 < y_2 < y_3$. This means that $0 < \text{Im}(v_0) < \text{Im}(\rho_1)$. The function $\wp$ is even and, hence, also $\tilde{v}_0 := 2\omega' - v_0 \in \mathbb{R}$ is a pole of second order (see Fig. 3.14).

Since $\wp$ is symmetric with respect to ρ_j, the function F_2 can be expanded in the same way as above as

$$F_2(v) = \frac{a_{j2}}{(v - \rho_j)^2} + \text{holomorphic part} \tag{A.2.13}$$

in a neighborhood of ρ_j. An expansion of $(2\wp(v) + \frac{1}{6})\wp'(v)$ near ρ_j yields

$$\left(2\wp(v) + \tfrac{1}{6}\right)\wp'(v) = \alpha_{j1}(v - \rho_j) + \alpha_{j2}(v - \rho_j)^2 + \text{higher orders} \tag{A.2.14}$$

because of $\wp'(\rho_j) = 0$. The coefficients are given by

$$\begin{aligned} \alpha_{j1} &= \left(\left(2\wp(v) + \tfrac{1}{6}\right)\wp'(v)\right)'\big|_{v=\rho_j} = \left(2y_j + \tfrac{1}{6}\right)\wp''(\rho_j) \\ \alpha_{j2} &= \left(\left(2\wp(v) + \tfrac{1}{6}\right)\wp'(v)\right)''\big|_{v=\rho_j} = 0. \end{aligned} \tag{A.2.15}$$

A comparison of coefficients

$$1 = F_2(v)\left(\left(2\wp(v) + \tfrac{1}{6}\right)\wp'(v)\right)^2 = a_{j2}\alpha_{j1}^2 + \text{higher powers of } (v - \rho_j) \tag{A.2.16}$$

yields

$$F_2(v) = \frac{1}{(v - \rho_j)^2}\left(\left(2y_j + \tfrac{1}{6}\right)\wp''(\rho_j)\right)^{-2} + \text{holomorphic part} \tag{A.2.17}$$

in a neighborhood of ρ_j.

The same procedure will be carried out for v_0 and $\tilde{v}_0$. We have

$$F_2(v) = \frac{b_2}{(v - v_0)^2} + \frac{b_1}{v - v_0} + \text{holomorphic part} \tag{A.2.18}$$

and

$$\left(2\wp(v) + \tfrac{1}{6}\right)\wp'(v) = \beta_1(v - v_0) + \beta_2(v - v_0)^2 + \text{higher orders} \tag{A.2.19}$$

near v_0. The coefficients of (A.2.19) read

$$\begin{aligned} \beta_1 &= 2\wp'(v_0)^2 \\ \beta_2 &= 3\wp'(v_0)\wp''(v_0). \end{aligned} \tag{A.2.20}$$

Again, a comparison of coefficients

$$\begin{aligned} 1 &= F_2(v)\left(\left(2\wp(v) + \tfrac{1}{6}\right)\wp'(v)\right)^2 \\ &= b_2\beta_1^2 + (2b_2\beta_1\beta_2 + b_1\beta_1^2)(v - v_0) + \text{higher orders} \end{aligned} \tag{A.2.21}$$

yields

$$\begin{aligned} b_2 &= \beta_1^{-2} = \frac{1}{4\wp'(v_0)^4} \\ b_1 &= -2\beta_1\beta_2 b_2 \beta_1^{-2} = -2\beta_2\beta_1^{-3} = -\frac{3}{4}\frac{\wp''(v_0)}{\wp'(v_0)^5}. \end{aligned} \tag{A.2.22}$$

In a neighborhood of $\tilde{v}_0$, the function F_2 is given by

$$F_2(v) = \frac{\tilde{b}_2}{(v - \tilde{v}_0)^2} + \frac{\tilde{b}_1}{v - \tilde{v}_0} + \text{holomorphic part}. \tag{A.2.23}$$

As $\wp'$ is an odd and $\wp''$ an even function we get for the coefficients of the expansion of $(2\wp(v)+\tfrac{1}{6})\wp'(v)$ near $\tilde{v}_0$ with $\tilde{v}_0 = 2\omega' - v_0$ the relations

$$\tilde{\beta}_1 = \beta_1, \quad \tilde{\beta}_2 = -\beta_2 \tag{A.2.24}$$

A. Elliptic integrals of third kind

and, therefore,
$$\tilde{b}_1 = -b_1, \quad \tilde{b}_2 = b_2. \tag{A.2.25}$$

Summarized, the function
$$g_2(v) := \sum_{j=1}^{3} a_{j2}\wp(v - \rho_j) + b_2(\wp(v - v_0) + \wp(v - \tilde{v}_0)) + b_1(\zeta(v - v_0) - \zeta(v - \tilde{v}_0)) \tag{A.2.26}$$

has the same poles with the same coefficients as F_2. Therefore, $F_2 - g_2$ is a holomorphic elliptic function and, thus, equal to a constant c_2 by Thm. 2.2. This constant can be determined by the condition $0 = F_2(0) = g_2(0) + c_2$ which yields
$$c_2 = -\sum_{j=1}^{3} a_{j2}y_j - \tfrac{1}{6}b_2 + b_1(-\zeta(v_0) + \zeta(\tilde{v}_0)). \tag{A.2.27}$$

As a consequence,
$$F_2(v) = \sum_{j=1}^{3} a_{j2}(\wp(v - \rho_j) - y_j) + b_2(\wp(v - v_0) + \wp(v - \tilde{v}_0) + \tfrac{1}{6})$$
$$+ b_1(\zeta(v - v_0) - \zeta(v - \tilde{v}_0) + \zeta(v_0) - \zeta(\tilde{v}_0)). \tag{A.2.28}$$

Now we can carry out the integration of the second term on the right hand side of (3.3.45):
$$\int_\gamma F_2(v)dv = \sum_{j=1}^{3} a_{j2}\left(\int_\gamma \wp(v - \rho_j)dv - 2\omega y_j\right) + b_2 \int_\gamma \wp(v - v_0) + \wp(v - \tilde{v}_0)dv$$
$$+ \tfrac{b_2}{3}\omega + b_1 \int_\gamma \zeta(v - v_0) - \zeta(v - \tilde{v}_0)dv + 2b_1\omega(\zeta(v_0) - \zeta(\tilde{v}_0))$$
$$= \sum_{j=1}^{3} -2a_{j2}(\eta + y_j\omega) + b_2\left(\tfrac{1}{3}\omega - 4\eta\right) + 2b_1\omega(\zeta(v_0) - \zeta(\tilde{v}_0))$$
$$+ b_1(2\eta(\tilde{v}_0 - v_0) + 2\pi i(k_1 - k_2))$$
$$= \sum_{j=1}^{3} -2a_{j2}(\eta + y_j\omega) + b_2\left(\tfrac{1}{3}\omega - 4\eta\right) + 2b_1\omega(2\zeta(v_0) - 2\eta')$$
$$+ b_1(2\eta(2\omega' - 2v_0) + 2\pi i(k_1 - k_2)). \tag{A.2.29}$$

The difference $(k_1 - k_2)$ can be calculated as follows. First note that via $u = 2\wp(v) + \tfrac{1}{6}$, v_0 corresponds to $u = 0$ and v_1 to $u = u_1$. Since $0 < u_1$ for the bound orbits under consideration we have $\text{Im}(v_0) < \text{Im}(v_1) < \text{Im}(\tilde{v}_0)$. Let now l be determined by
$$\int_{v_2}^{v_2+2\omega} \zeta(v - \tilde{v}_0)dv = 2\eta(v_2 - \tilde{v}_0 + \omega) + \pi i + 2\pi i l, \tag{A.2.30}$$
where $v_2 \in i \cdot \mathbb{R}$ is such that $\text{Im}(v_2) > \text{Im}(\tilde{v}_0) > \text{Im}(v_1) > \text{Im}(v_0)$. From

(i) l does not depend on $\tilde{v}_0$ as long as $\text{Im}(\tilde{v}_0) < \text{Im}(v_2)$ holds and, thus,

$$\int_{v_2}^{v_2+2\omega} \zeta(v-v_0)dv = 2\eta(v_2 - v_0 + \omega) + \pi i + 2\pi i l \tag{A.2.31}$$

and

(ii) (A.2.31) holds also for v_2 replaced by v_1 by Cauchy's integral formula for the rectangle with corners v_1, $v_1 + 2\omega$, $v_2 + 2\omega$, and v_2

it follows that $l = k_1$.

We show now that $k_2 = l + 1$ and, thus, $k_2 = k_1 + 1$. Consider the counterclockwise oriented rectangle with corners v_1, $v_1 + 2\omega$, $v_2 + 2\omega$, and v_2. Let c be the boundary of this rectangle but with a two symmetric small bumps such that c encircles the pole $\tilde{v}_0$ of $\zeta(v - \tilde{v}_0)$ with residue 1, but not $\tilde{v}_0 + 2\omega$. Then the residue theorem gives

$$\begin{aligned}
2\pi i &= \oint_c \zeta(v - \tilde{v}_0)dv \\
&= \int_{v_1}^{v_1+2\omega} \zeta(v - \tilde{v}_0)dv + \int_{v_1+2\omega}^{v_2+2\omega} \zeta(v - \tilde{v}_0)dv + \int_{v_2+2\omega}^{v_2} \zeta(v - \tilde{v}_0)dv \\
&\quad + \int_{v_2}^{v_1} \zeta(v - \tilde{v}_0)dv \\
&= \int_{v_1}^{v_1+2\omega} \zeta(v - \tilde{v}_0)dv - \int_{v_2}^{v_2+2\omega} \zeta(v - \tilde{v}_0)dv + 2\eta(v_2 - v_1) \\
&= 2\eta(v_1 - \tilde{v}_0 + \omega) + \pi i + 2\pi i k_2 - (2\eta(v_2 - \tilde{v}_0 + \omega) + \pi i + 2\pi i l) + 2\eta(v_2 - v_1) \\
&= 2\pi i(k_2 - l).
\end{aligned} \tag{A.2.32}$$

With the Legendre relation $4(\eta\omega' - \eta'\omega) = 2\pi i$ we finally obtain

$$\int_\gamma F_2(v)dv = \sum_{j=1}^{3} \frac{-2(\eta + y_j\omega)}{(2y_j + \frac{1}{6})^2 \wp''(\rho_j)^2} + \frac{\omega - 12\eta}{12\wp'(v_0)^4} + 3\frac{\wp''(v_0)}{\wp'(v_0)^5}(\eta v_0 - \omega\zeta(v_0)). \tag{A.2.33}$$

Note that though the values $\wp'(v_0)^5$, v_0 and $\zeta(v_0)$ appearing in the last part of the right-hand side are all purely imaginary, the hole term is real.

APPENDIX B

Calculational details

In this appendix some details on the explicite calculation of solutions of geodesic equations are given. This includes useful rearrangements of typical integral expressions, the calculation of periods for elliptic as well as hyperelliptic functions, and the description of a Newton method which can be used to determine the redundant parameter appearing in solutions with hyperelliptic functions.

B.1 Typical integral expressions

In this section, the calculation of integrals of the type $\int_{x_1}^{x_2} \frac{x^k dx}{\sqrt{P_l(x)}}$ will be discussed, where $P_l(x) = c\prod_{j=1}^{l}(x - e_j)$ with a constant c, $l = 3$ or $l = 5$, and $k = 0, \ldots, \frac{l-3}{2}$. This kind of integral appears frequently in the calculation of solutions of geodesic equations for the cases that (i) x_1 and x_2 are real (and maybe equal to a real zero of P_l) or that (ii) x_1 and x_2 are two conjugate complex zeros of P_l. Two difficulties may arise in the calculation. The first is the the occurrence of a singularity if one of the zeros of P_l is equal to x_1 or x_2 and the second the control of the branch of the square root.

Assume now that the integration path is real. In this case, concerning the second point, it is helpful to rearrange P_l such that $P_l(x) = \pm c\bar{P}_l(x)$ with $\bar{P}_l(x) \geq 0$ on the integration path (if the sign of P_l changes on the integration path, it should be divided such that on each part either $P_l \geq 0$ or $P_l \leq 0$). Then $\sqrt{P_l(x)} = \sqrt{\pm c}\sqrt{\bar{P}_l(x)}$, where on the right hand side the second square root is positive and the first has the same branch as the left hand side. The constant factor $\sqrt{\pm c}$ can then be extracted

B.1. Typical integral expressions

from the integral and dealt with separately. If now x_1 coincides with the real zero e_i of $\bar{P}_l(x)$ this singularity can be eliminated by an integration by parts

$$\int_{e_i}^{x_2} \frac{x^k dx}{\sqrt{\bar{P}_l(x)}} = \left(\frac{2\sqrt{x-e_i}\,x^k}{\sqrt{Q_1(x)}}\right)\bigg|_{e_i}^{x_2} - 2k\int_{e_i}^{x_2} \frac{\sqrt{x-e_i}\,x^k}{\sqrt{Q_1(x)}}dx + \int_{e_i}^{x_2} \frac{\sqrt{x-e_i}\,x^k}{\sqrt{Q_1(x)}^3} Q_1'(x)dx$$

$$= \frac{2\sqrt{x_2-e_i}\,x_2^k}{\sqrt{Q_1(x_2)}} - 2k\int_{e_i}^{x_2} \frac{\sqrt{x-e_i}}{\sqrt{Q_1(x)}}dx + \int_{e_i}^{x_2} \frac{\sqrt{x-e_i}\,x^k}{\sqrt{Q_1(x)}^3} Q_1'(x)dx, \quad (B.1.1)$$

where $Q_1(x) = \frac{\bar{P}_l(x)}{(x-e_i)} = \pm\prod_{i\neq j}(x-e_j) \geq 0$ on $[e_i, x_2]$. Analogously the case of $x_2 = e_i$ can be treated yielding

$$\int_{x_1}^{e_i} \frac{x^k dx}{\sqrt{\bar{P}_l(x)}} = -\frac{2\sqrt{e_i-x_1}\,x_1^k}{\sqrt{Q_2(x_1)}} + 2k\int_{x_1}^{e_i} \frac{\sqrt{e_i-x}\,x^k}{\sqrt{Q_2(x)}}dx - \int_{x_1}^{e_i} \frac{\sqrt{e_i-x}\,x^k}{\sqrt{Q_2(x)}^3} Q_2'(x)dx, \quad (B.1.2)$$

where $(e_i - x)Q_2(x) = \bar{P}_l(x)$. If both $x_1 = e_i$ and $x_2 = e_{i+1}$ the integration path can be divided somewhere in between and both formulas have to be applied. Note that for all integrations the square root is always positive.

In a similar way the case of x_1 and x_2 being two conjugate complex zero of P_l can be handled. First divide the integration path at $p := \mathrm{Re}(x_1)$ in two halves. The new integration path can then be parameterized by $x = p + It$, $t \in [0, \mathrm{Im}(x_2)]$, ($I$ is the imaginary unit) and $P_l(x)$ expressed by

$$P_l(p + It) = (\mathrm{Im}(x_2) - t)(t + \mathrm{Im}(x_2))Q_3(t), \quad (B.1.3)$$

where $Q_3(t) = P_l(p+It)(p+It-e_i)^{-1}(p+It-e_{i+1})^{-1}$ assuming $x_1 = e_i, x_2 = e_{i+1}$. (Here we used the fact that $\mathrm{Re}(x_1) = p = \mathrm{Re}(x_2)$ and $\mathrm{Im}(x_1) = -\mathrm{Im}(x_2)$.) Then an integration by parts yields

$$\int_p^{x_2} \frac{dx}{\sqrt{P_l(x)}} = I\int_0^q \frac{dt}{\sqrt{P_l(p+It)}}$$

$$= I\left[\frac{2}{\sqrt{Q_3(0)}} - \int_0^q \frac{\sqrt{q-t}}{\sqrt{t+q}^3}\frac{1}{\sqrt{Q_3(t)}}dt - \int_0^q \frac{\sqrt{q-t}}{\sqrt{t+q}}\frac{Q_3'(t)}{\sqrt{Q_3(t)}^3}dt\right] \quad (B.1.4)$$

where $q = \mathrm{Im}(x_2)$. For $k = 1$ we get analogously

$$\int_p^{x_2} \frac{x\,dx}{\sqrt{P_l(x)}} = -\int_0^q \frac{t\,dt}{\sqrt{P_l(p+It)}} + Ip\int_0^q \frac{dt}{\sqrt{P_l(p+It)}}, \quad (B.1.5)$$

$$\int_0^q \frac{t\,dt}{\sqrt{P_l(p+It)}} = 2\int_0^q \frac{\sqrt{q-t}}{\sqrt{t+q}}\frac{1}{\sqrt{Q_3(t)}}dt - \int_0^q \frac{\sqrt{q-t}}{\sqrt{t+q}^3}\frac{t}{\sqrt{Q_3(t)}}dt - \int_0^q \frac{\sqrt{q-t}}{\sqrt{t+q}}\frac{tQ_3'(t)}{\sqrt{Q_3(t)}^3}dt.$$
$$(B.1.6)$$

The square roots of $(q+t)$ and $(q-t)$ are always positive whereas the evaluation of the square root of $Q_3(t) \in \mathbb{C}$ has to be explicitly controlled. This can be done by using

$$\sqrt{Q(t)} = \exp\left[-\frac{1}{2}\log(\mathrm{Re}(Q(t)) + I\mathrm{Im}(Q(t)))\right]\exp(-\pi In)$$

$$= \exp\left[-\frac{1}{2}(\ln|Q(t)| + i\arg(Q(t)))\right]\exp(-\pi In), \quad (B.1.7)$$

B. Calculational details

where $-\pi \leq \arg(Q(t)) < \pi$ is the argument of $Q(t)$ and $n \in \mathbb{Z}$.

B.2 Calculation of periods

The elliptic and hyperelliptic functions used in this book all depend on the period matrix corresponding to the underlying problem. In principle, three different types of period matrices are needed: the period matrix of first kind for a Riemann surface of genus one, the period matrix of first kind for a Riemann surface of genus two, and the period matrix of second kind for a Riemann surface of genus two. The period matrices of first kind are computed by integrating the vector of holomorphic differentials $d\vec{z} = (z^{i-1}/\sqrt{P(z)})_{i=1,\ldots,g}^t$, where g is the genus, along the integration paths given by the homology basis of the Riemann surface. The periods of second kind are computed by integrating the vector of meromorphic differentials along the same paths (see also [63]).

Therefore, the actual computation of a period matrix directly depends on the integration paths which in turn depend on the branch cuts. The paths around a branch cut are called a-paths and the paths from one branch cut to another and way back on the other sheet are called b-paths. A canonical choice is one where the paths a_i are disjoint to each other as are the paths b_i, and a_i and b_j have one common point if $i = j$ and else are also disjoint. For the case that the polynomial P defining the Riemann surface $y^2 = P(x)$ has only real zeros a canonical choice of branch cuts and paths is shown in Fig. 2.1. If P has also complex zeros, the branch cuts can be arranged in seven basically different ways. The branch cuts and paths chosen for all possible arrangements of zeros of P for the case of genus two together with the appropriate sign of the square root on one sheet (the sign is reversed on the other sheet) are shown in Fig. B.1. For the case of genus one, there are only three cases corresponding to simply neglecting e_1 and e_2 in Figs. B.1(a), (d), and (e). So far, the periods have to be calculated by numerical integration. However, it may also be possible to express them in terms of zeros of the theta function, cf. [88].

Once a choice of branch cuts and paths is done, we can integrate the holomorphic and meromorphic differentials. For paths which go around real zeros e_i, e_{i+1} of P, this integral can be computed as two times the integral from e_i to e_{i+1} along the real axis, because the integration back from e_{i+1} to e_i yields just the same value due to the changed sign of the square root. For paths which go around one or two complex zeros, the integral can be computed along a path running from the complex zero to its real part and onward to the other zero. Note that this path is of course not allowed to cross another branch cut. This means that for example in the case B.1(c) the period corresponding to the a_1-path is computed along a path starting from e_3 running to the real part of e_3 onward to e_1 and back to the real part of e_3 - this does not cancel due to changed signs of the square root - and then finishing at e_2. The resulting period matrix can be checked by the Legendre relation (2.2.5) and by the properties of the

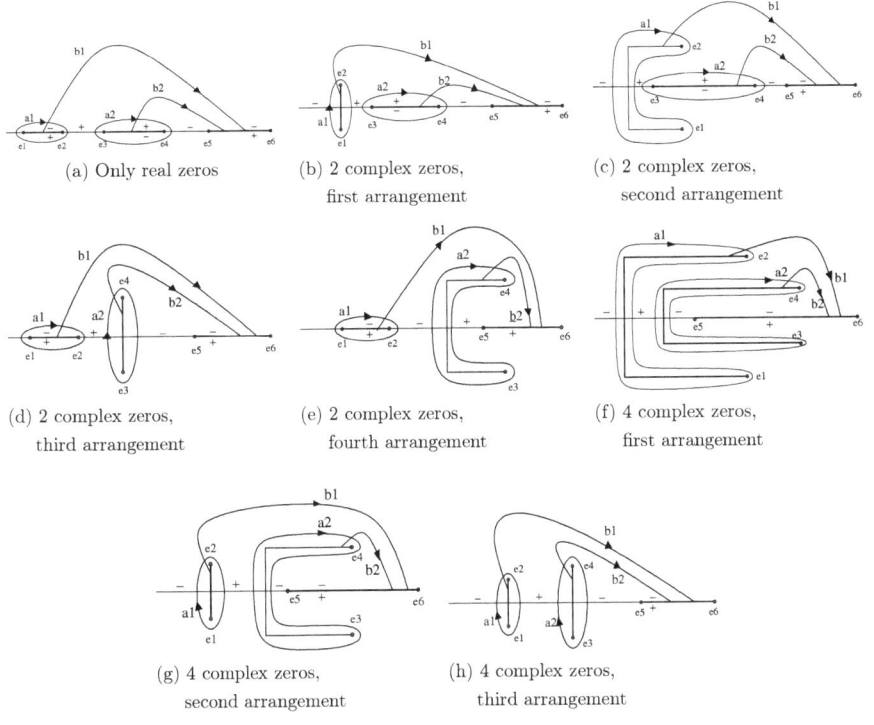

Figure B.1: Branch cuts and homology basis of paths for all arrangements of zeros $e_1, \ldots, e_5$ of a polynomial of degree 5. The b-paths are completed on the other sheet.

Riemann matrix τ, i.e. symmetry and positive definite imaginary part.

B.3 Calculation of redundant parameter

For solutions in terms of derivatives of Kleinian σ functions elements of the theta divisor has to calculated. As the theta divisor is a one-dimensional submanifold of the Jacobian, one of its two components can be calculated from the other by taking into account that the Kleinian σ function vanishes on the theta divisor. The physical parameter $s - s_{\text{in}}$, which is one of the components of the theta divisor, is given by $s = \gamma$ with the Mino time γ for axial symmetric problems and by $s = \varphi$ with the angle φ for spherically symmetric problems. The constant s_{in} depends on the initial values s_0 and x_0. The calculation of the unphysical redundant parameter can then be done by applying a Newton

B. Calculational details

method to the function $f = g \circ h : \mathbb{C} \to \mathbb{C}^2 \to \mathbb{C}$,

$$w \overset{h}{\mapsto} z \overset{g}{\mapsto} \vartheta \left[\begin{pmatrix} 1/2 \\ 1/2 \end{pmatrix}, \begin{pmatrix} 0 \\ 1/2 \end{pmatrix} \right] (z; \tau), \tag{B.3.1}$$

where $z = (2\omega)^{-1} \left(\frac{s - s_{in}}{w} \right)$ with the period matrix ω if the second component of the theta divisor is redundant and $z = (2\omega)^{-1} \left(\frac{w}{s - s_{in}} \right)$ if the first is redundant. Since f is a complex function it has to be interpreted as a mapping from $\mathbb{R}^2 \to \mathbb{R}^4 \to \mathbb{R}^2$ for the purpose of computing w. For each Newton iteration $w_{n+1} = w_n - J(f(w_n))^{-1} f(w_n)$ the Jacobi matrix $J(f) = J(g) \cdot J(h)$ and, hence, the derivatives of $\text{Re}(g) = \text{Re}\left(\vartheta\left[\begin{pmatrix} 1/2 \\ 1/2 \end{pmatrix}, \begin{pmatrix} 0 \\ 1/2 \end{pmatrix}\right]\right)$ and $\text{Im}(g)$ are needed. They are given by

$$\begin{aligned}
\frac{\partial \text{Re}(g)}{\partial \text{Re}(z_i)} =& -2\pi \sum_{|m|=-\infty}^{\infty} e^{\pi \left(m + \begin{pmatrix} 1/2 \\ 1/2 \end{pmatrix} \right)^t \left[\text{Re}(i\tau)\left(m + \begin{pmatrix} 1/2 \\ 1/2 \end{pmatrix}\right) - 2\text{Im}(z) \right]} \left(m_i + \begin{pmatrix} 1/2 \\ 1/2 \end{pmatrix} \right) \\
& \sin\left(\pi \left(m + \begin{pmatrix} 1/2 \\ 1/2 \end{pmatrix} \right)^t \left[\text{Im}(i\tau)\left(m + \begin{pmatrix} 1/2 \\ 1/2 \end{pmatrix}\right) + 2\text{Re}(z) + 2\begin{pmatrix} 0 \\ 1/2 \end{pmatrix} \right] \right), \\
\frac{\partial \text{Im}(g)}{\partial \text{Re}(z_i)} =& 2\pi \sum_{|m|=-\infty}^{\infty} e^{\pi \left(m + \begin{pmatrix} 1/2 \\ 1/2 \end{pmatrix} \right)^t \left[\text{Re}(i\tau)\left(m + \begin{pmatrix} 1/2 \\ 1/2 \end{pmatrix}\right) - 2\text{Im}(z) \right]} \left(m_i + \begin{pmatrix} 1/2 \\ 1/2 \end{pmatrix} \right) \\
& \cos\left(\pi \left(m + \begin{pmatrix} 1/2 \\ 1/2 \end{pmatrix} \right)^t \left[\text{Im}(i\tau)\left(m + \begin{pmatrix} 1/2 \\ 1/2 \end{pmatrix}\right) + 2\text{Re}(z) + 2\begin{pmatrix} 0 \\ 1/2 \end{pmatrix} \right] \right).
\end{aligned} \tag{B.3.2}$$

The derivatives with respect to $\text{Im}(z_i)$ can be derived from the Cauchy-Riemann differential equation and need not to be computed. To calculate $x(s)$ for a list of values $s = s_1, \ldots, s_m$ the starting value for the Newton iteration to compute w corresponding to s_i can be chosen as the final w corresponding to s_{i-1}. Once the Newton iteration successfully found a zero of f, this zero can be inserted into the formula of the solution of the geodesic equation.

Bibliography

[1] A. Einstein. Explanation of the perihelion motion of Mercury from the general theory of relativity. *Sitzungsber. Preuss. Akad. Wiss.*, (Part 2):831–839, 1915.

[2] K. Schwarzschild. Über das Gravitationsfeld eines Massenpunktes nach der Einstein'schen Theorie. *Sitzungsber. Preuss. Akad. Wiss.*, page 189, 1916.

[3] F.W. Dyson, A.S. Eddington, and C. Davidson. A determination of the deflection of light by the Sun's gravitational field, from observations made at the total eclipse of May 29, 1919. *Philos. Trans. Royal Soc. London*, 220A:291, 1919.

[4] J. Ehlers, F.A.E. Pirani, and A. Schild. The geometry of free fall and light propagation. In L. O'Raifeartaigh, editor, *General Relativity, Papers in Honour of J.L. Synge*, page 63. Clarendon Press, 1972.

[5] J. Ehlers. Survey of General Relativity Theory. In W. Israel, editor, *Relativity, Astrophysics and Cosmology*, page 1. Reidel, 1973.

[6] Y. Hagihara. Theory of relativistic trajectories in a gravitational field of Schwarzschild. *Japan. J. Astron. Geophys.*, 8:67, 1931.

[7] H. Reissner. Über die Eigengravitation des elektrischen Feldes nach der Einsteinschen Theorie. *Annalen der Physik*, 50:106, 1916.

[8] G. Nordström. On the energy of the gravitational field in Einstein's theory. *Verhandl. Koninkl. Ned. Akad. Wetenschap., Afdel. Natuurk*, 26:1201.

Bibliography

[9] J. Lense and H. Thirring. Über den Einfluß der Eigenrotation der Zentralkörper auf die Bewegung der Planeten und Monde nach der Einsteinschen Gravitations-theorie. *Phys. Zeitschrift*, 19:156, 1918.

[10] R.P. Kerr. Gravitational field of a spinning mass as an example of algebraically special metrics. *Phys. Rev. Let.*, 11:237, 1963.

[11] B. Carter. Global structure of the Kerr family of gravitational fields. *Phys. Rev.*, 174, 5:1559, 1968.

[12] B. Carter. Black hole equilibrium states. In C. deWitt and B. deWitt, editors, *Black Holes - Les astres occlus*, page 61. Gordon and Breach Science Publishers, New York-London-Paris, 1973.

[13] F. de Felice. Equatorial geodesic motion in gravitational field of a rotating source. *Nuovo Cimento B*, 57(2):351, 1968.

[14] S. Chandrasekhar. *The Mathematical Theory of Black Holes*. Oxford University Press, Oxford, 1983.

[15] G. Slezáková. *Geodesic geometry of black holes*. Phd thesis, University of Waikato, Waikato, New Zealand, 2006.

[16] A. Einstein. Cosmological observations on the general theory of relativity. *Sitzungsber. Preuss. Akad. Wiss.*, (Part 1):142–152, 1917.

[17] A.G. Riess et al. Observational Evidence from Supernovae for an Accelerating Universe and a Cosmological Constant. *Astronom. J.*, 116:1009, 1998.

[18] S. Perlmutter et al. Measurements of omega and lambda from 42 high-redshift supernovae. *Astrophys. J.*, 517:565, 1999.

[19] C. L. Bennett et al. First-year Wilkinson Microwave Anisotropy Probe (WMAP) observations: Preliminary maps and basic results. *Astrophys. J. Suppl. Ser.*, 148:1, 2003.

[20] D. N. Spergel et al. First-year Wilkinson Microwave Anisotropy Probe (WMAP) observations: Determination of cosmological parameters. *Astrophys. J. Suppl. Ser.*, 148:175, 2003.

[21] I. Zlatev, L. Wang, and Steinhardt P.J. Quintessence, cosmic coincidence, and the cosmological constant. *Phys. Rev. Lett.*, 82(5):896, 1999.

[22] P. J. Steinhardt, L. Wang, and I. Zlatev. Cosmological tracking solutions. *Phys. Rev. D*, 59:123504, 1999.

[23] V. Kagramanova, J. Kunz, and C. Lämmerzahl. Solar system effects in Schwarzschild–de Sitter space–time. *Phys. Lett.*, B 634:465, 2006.

[24] P. Jetzer and M. Sereno. Two-body problem with the cosmological constant and observational constraints. *Phys. Rev.*, D 73:044015, 2006.

[25] A.W. Kerr, J.C. Hauck, and B. Mashhoon. Standard clocks, orbital precession and the cosmological constant. *Class. Qauntum Grav.*, 20:2727, 2003.

[26] J.D. Anderson et al. Study of the anomalous acceleration of Pioneer 10 and 11. *Phys. Rev.*, D 65:082004, 2002.

[27] J. Näf, P. Jetzer, and M. Sereno. On gravitational waves in spacetimes with a nonvanishing cosmological constant. *Phys. Rev.*, D 79:024014, 2009.

[28] C. Barabèz and P.M. Hogan. Inhomogeneous high frequency expansion-free gravitational waves. *Phys. Rev.*, D 75:124012, 2007.

[29] J. Dexter and E. Agol. A fast new public code for computing photon orbits in a Kerr spacetime. *Astrophys. J.*, 696:1616, 2009.

[30] Y. Mino. Perturbative approach to an orbital evolution around a supermassive black hole. *Phys. Rev.*, D 67:084027, 2003.

[31] I.S. Gradshteyn and I.M. Ryzhik. *Table of Integrals, Series, and Products.* Academic Press, Orlando, 1983.

[32] C.G.J. Jacobi. *Gesammelte Werke.* Reimer, Berlin, 1881.

[33] N.H. Abel. *Oeuvres complètes de Niels Henrik Abel.* 1881.

[34] B. Riemann. Theorie der Abel'schen Functionen. *Crelle's J.*, 54:115, 1857.

[35] B. Riemann. Über das Verschwinden der ϑ-Functionen. *Crelle's J.*, 65:161, 1866.

[36] K.T.W. Weierstrass. Zur Theorie der Abelschen Functionen. *Crelle's Journal*, 47:289, 1854.

[37] H.F. Baker. *Abelian Functions. Abel's theorem and the allied theory of theta functions.* Cambridge University Press, Cambridge, 1995. First published 1897.

[38] S. Matsutani. Hyperelliptic solutions of KdV and KP equations: Reevaluation of Baker's study on hyperelliptic sigma functions. *arXiv*, nlin.SI/00070001, 2000.

[39] J. C. Eilbeck, V. Z. Enolskii, and N. A. Kostov. Quasi periodic solutions for vector nonlinear Schrödinger equations. *J. Math. Phys.*, 41:8236, 2000.

[40] G.V. Kraniotis and S.B. Whitehouse. Compact calculation of the perihelion precession of Mercury in general relativity, the cosmological constant and Jacobis inversion problem. *Class. Quantum Grav.*, 20:4817, 2003.

[41] G.V. Kraniotis. Precise relativistic orbits in Kerr and Kerr–(anti) de Sitter spacetimes. *Class. Quantum Grav.*, 21:4743, 2004.

[42] R. Fujita and W. Hikida. Analytical solutions of bound timelike geodesic orbits in Kerr spacetime. *Class. Quantum Grav.*, 26:135002, 2009.

[43] A. Hurwitz. *Vorlesungen über Allgemeine Funktionentheorie und elliptische Funktionen.* Springer–Verlag, Berlin, 1964.

[44] A.I. Markushevich. *Theory of complex function 1-3.* Chelsea Pub. Co., New York, 1977.

[45] W. Fischer and I. Lieb. *Funktionentheorie.* Vieweg, 8. edition, 2003.

[46] R. Miranda. *Algebraic Curves and Riemann Surfaces.* American Math. Soc., Providence, 1995.

[47] H.E. Rauch and H.M. Farkas. *Theta Functions with Applications to Riemann Surfaces.* Williams and Wilkins, Baltimore, 1974.

[48] V.M. Buchstaber, V.Z. Enolskii, and D.V. Leykin. *Hyperelliptic Kleinian Functions and Applications.* Reviews in Mathematics and Mathematical Physics 10. Gordon and Breach, 1997.

[49] D. Mumford. *Tata Lectures on Theta, Vol. I and II.* Birkhäuser, Boston, 1983/84.

[50] H.F. Baker. *Multiply Periodic Functions.* Cambridge University Press, 1907.

[51] V.Z. Enolskii, M. Pronine, and P.H. Richter. Double pendulum and θ-divisor. *J. Nonlinear Sc.*, 13:157, 2003.

[52] S. Abenda and Yu Fedorov. On the weak Kowalevski-Painlevé property for hyperelliptic separable systems. *Acta Appl. Math.*, 60:137, 2000.

[53] E. Hackmann, V. Kagramanova, J. Kunz, and C. Lämmerzahl. Analytic solutions of the geodesic equations in higher dimensional static spherically symmetric space–times. *Phys. Rev.*, D 78:124018, 2008.

[54] F. Kottler. Über die physikalischen Grundlagen der Einsteinschen Gravitationstheorie. *Ann. Phys. (Leipzig)*, 361:401, 1918.

[55] K.H. Geyer. Geometrie der Raum-Zeit der Maßbestimmung von Kottler, Weyl und Trefftz. *Astron. Nachr.*, 301:135, 1980.

[56] F.R. Tangherlini. Schwarzschild field in n dimensions and the dimensionality of space problem. *Nuovo Cim.*, 27:636, 1963.

[57] V. Kagramanova. *Motion in General Relativity. Investigation of spherically and axially symmetric spacetimes through geodesics.* Phd thesis, Carl von Ossietzky Universität Oldenburg, 2009.

[58] C.W. Misner, K. Thorne, and J.A. Wheeler. *Gravitation.* Freeman, San Francisco, 1973.

[59] H.J. Schmidt. On the de Sitter space-time - the geometric foundation of inflationary cosmology. *Fortschr. Phys.*, 41:179, 1993.

[60] P.J.E. Peebles and B. Ratra. The cosmological constant and dark energy. *Rev. Mod. Phys.*, 75:559, 2003.

[61] E. Hackmann and C. Lämmerzahl. Complete analytic solution of the geodesic equation in Schwarzschild–(anti) de Sitter space–times. *Phys. Rev. Lett.*, 100:171101–1, 2008.

[62] E. Hackmann and C. Lämmerzahl. Geodesic equation in Schwarzschild-(anti-)de Sitter spacetimes: Analytical solutions and applications. *Phys. Rev.*, D 78:024035, 2008.

[63] E. Hackmann and C. Lämmerzahl. Geodesic equation and theta-divisor. In *Recent Developments in Gravitation and Cosmology*, volume 977 of *AIP Conference Proceedings*, page 116, 2008.

[64] E. Hackmann and C. Lämmerzahl. Hyperelliptic functions and geodesic equations. *Proceedings in Applied Mathematics and Mechanics*, 8:10723, 2008.

[65] W. Rindler and M. Ishak. Contribution of the cosmological constant to the relativistic bending of light revisited. *Phys. Rev.*, D 76:043006, 2007.

[66] M.J. Valtonen. OJ287: A binary black hole system. *Rev. Mex. Astron. Astrofis. Conf. ser.*, 32:22, 2008.

[67] M.J. Valtonen et al. A massive binary black hole system in OJ287 and a test of general relativity. *Nature*, 452:851, 2008.

[68] B. Rievers et al. New powerful thermal modelling for high-precision gravity missions with application to Pioneer 10/11. *New J. Phys.*, 11:113032, 2009.

[69] B. Rievers et al. Thermal dissipation force modeling with preliminary results for Pioneer 10/11. *Acta Astronautica*, 66:467, 2010.

[70] M.M. Nieto and J.D. Anderson. Using early data to illuminate the Pioneer anomaly. *Class. Quantum. Grav.*, 22:5343, 2005.

[71] S. W. Hawking and G.F.R. Ellis. *The large scale structure of space-time*. Cambridge Univ. P., Cambridge, 1973.

[72] M. Demianski and M. Francaviglia. Separability structures and Killing-Yano tensors in vacuum type-D space-times without acceleration. *Int. J. Theor. Phys.*, 19:675, 1980.

[73] R. Debever, N. Kamran, and R.G. McLenaghan. Exhaustive integration and a single expression for the general-solution of the type-D vacuum and electrovac field-equations with cosmological constant for a nonsingular aligned Maxwell field. *J. Math. Phys.*, 25:1955, 1984.

[74] D. Kubizňák and P. Krtouš. Conformal Killing-Yano tensors for the Plebański-Demiański family of solutions. *Phys. Rev.*, D 76:084036, 2007.

[75] B. O'Neill. *The Geometry of Black holes*. A K Peters, Wellesly, Massasuchetts, 1995.

[76] J. Levin and G. Perez-Giz. Homoclinic orbits around spinning black holes. I. Exact solution for the Kerr separatrix. *Phys. Rev.*, D 79:124013, 2009.

[77] E. Hackmann, V. Kagramanova, J. Kunz, and C. Lämmerzahl. Analytic solutions of the geodesic equation in axially symmetric space–times. *Europhys. Lett.*, 88:30008, 2009.

[78] E. Hackmann, Kagramanova K., J. Kunz, and C. Lämmerzahl. Analytical solution of the geodesic equation in Kerr-(anti) de Sitter space-time. *Phys. Rev.*, D 81:044020, 2010.

[79] C.M. Will. *Theory and Experiment in Gravitational Physics* . Cambridge University Press, Cambridge, revised edition, 1993.

[80] I. Ciufolini. Dragging of inertial frames. *Nature*, 449:41, 2007.

[81] L. Schiff. Possible new experimental test of general relativity theory. *Phys. Rev. Lett.*, 4:215, 1960.

[82] F. Everitt et al. Gravity Probe B data analysis - Status and potential for improved accuracy of scientific results. *Space Sci. Rev.*, 148:53, 2009.

[83] G. Schäfer. Gravitomagnetic Effects. *Gen.Rel.Grav.*, 36:2223, 2004.

[84] G. Schäfer. Gravitomagnetism in Physics and Astrophysics. *Space Sci. Rev.*, 148:37, 2009.

[85] S. Drasco and S.A. Hughes. Rotating black hole orbit functionals in the frequency domain. *Phys. Rev.*, D 69:044015, 2004.

[86] J.F. Plebański and M. Demiański. Rotating, charged, and uniformly accelerating mass in general relativity. *Ann. Phys.*, 98:90, 1976.

[87] J.B. Griffiths and J. Podolsky. A new look at the Plebanski-Demianski family of solutions. *Int. J. Mod. Phys.*, 15:335, 2006.

[88] V.Z. Enolskii and P.H. Richter. Periods of hyperelliptic integrals expressed in terms of theta-constants by means of Thomae formulae. *Philos. Trans. R. Soc. A*, 366:1005, 2008.

[89] E. Hackmann, C. Lämmerzahl, and A. Macias. Complete classification of geodesic motion in fast Kerr and Kerr-(anti-)de Sitter space-times. In *New trends in statistical physics: Festschrift in honour of Leopoldo Garcia-Colin's 80 birthday*. World Scientific, Singapore, 2010, to appear.

[90] Y. Hagihara. *Celestial Mechanics*. MIT Press, Cambridge, Mass., 1970.

[91] E. Hackmann et al. The complete set of solutions of the geodesic equations in the space-time of a Schwarzschild black hole pierced by a cosmic string. *Phys. Rev.*, D 81:064016, 2010.

[92] V. Kagramanova, J. Kunz, and C. Lämmerzahl. Charged particle interferometry in Plebański-Demiański black hole spacetimes. *Class. Quant. Grav.*, 25:105023, 2008.

[93] V. Kagramanova, J. Kunz, E. Hackmann, and C. Lämmerzahl. Analytic treatment of complete and incomplete geodesics in Taub-NUT space-times. *Phys. Rev.*, D 81:124044, 2010.

[94] V. Kagramanova, J. Kunz, and C. Lämmerzahl. Analytic solutions of the geodesic equations in Taub-NUT-de Sitter spacetimes. In preparation.

[95] T. Damour. Coalescence of two spinning black holes: An effective one-body approach. *Phys.Rev.*, D 64:124013, 2001.

[96] T. Damour, P. Jaranowski, and G. Schäfer. Hamiltonian of two spinning compact bodies with next-to-leading order gravitational spin-orbit coupling. *Phys. Rev.*, D 77:064032, 2008.

[97] T. Damour, P. Jaranowski, and G. Schäfer. Effective one body approach to the dynamics of two spinning black holes with next-to-leading order spin-orbit coupling. *Phys.Rev.*, D78:024009, 2008.

[98] H. Quevedo. Multipole moments in General Relativity - Static and stationary vacuum solutions. *Fortschr. Phys.*, 38:733, 1990.

[99] H. Quevedo. General static axisymmetric solution of Einstein's vacuum field equations in prolate spheroidal coordinates. *Phys. Rev.*, D 39(10):2904, 1989.

Die VDM Verlagsservicegesellschaft sucht für wissenschaftliche Verlage abgeschlossene und herausragende

Dissertationen, Habilitationen, Diplomarbeiten, Master Theses, Magisterarbeiten usw.

für die kostenlose Publikation als Fachbuch.

Sie verfügen über eine Arbeit, die hohen inhaltlichen und formalen Ansprüchen genügt, und haben Interesse an einer honorarvergüteten Publikation?

Dann senden Sie bitte erste Informationen über sich und Ihre Arbeit per Email an *info@vdm-vsg.de*.

Sie erhalten kurzfristig unser Feedback!

VDM Verlagsservicegesellschaft mbH
Dudweiler Landstr. 99   Telefon +49 681 3720 174
D - 66123 Saarbrücken   Fax   +49 681 3720 1749
www.vdm-vsg.de

Die VDM Verlagsservicegesellschaft mbH vertritt

Printed by Books on Demand GmbH, Norderstedt / Germany